A SNORKELLER'S GUIDE TO THE
MEDITERRANEAN

A SNORKELLER'S GUIDE TO THE
MEDITERRANEAN

Keith Broomfield

PELAGIC PUBLISHING

Pelagic Publishing
20–22 Wenlock Road
London N1 7GU, UK

www.pelagicpublishing.com

A Snorkeller's Guide to the Mediterranean

https://doi.org/10.53061/XNUO6313

A CiP record for this book is available from the British Library

ISBN 978-1-78427-486-3 Pbk
ISBN 978-1-78427-487-0 ePub
ISBN 978-1-78427-488-7 PDF

Cover background photograph: Seaphotoart / Alamy Stock Photo
All other cover images © the author
Frontispiece: A shoal of Salema near a reef
All interior photographs by the author unless otherwise credited
Map of the Mediterranean and fish diagram created by Chris Bromley

Typeset in Minion Pro by S4Carlisle Publishing Services, Chennai, India

MIX
Paper | Supporting
responsible forestry
FSC® C014540

Printed by Short Run Press Ltd, Exeter

CONTENTS

PREFACE

I have been fascinated by Mediterranean marine life ever since having first donned a snorkel and mask on the Greek island of Rhodes in 1980. I still recall with clarity the glittering shoals of fish and the wonderfully warm turquoise water.

The experience led to me to purchasing *Fishes of the Sea* by John and Gillian Lythgoe (Blandford Press, 1971), and the book became my constant companion on every subsequent trip for many years to the Mediterranean, the pages well thumbed and providing an essential identification guide for my snorkelling adventures.

That book also catalysed a yearning to write my own identification guide to marine life for people visiting the Mediterranean who enjoy snorkelling and are interested in nature. *A Snorkeller's Guide to the Mediterranean* is the end result. I hope readers find it to be a useful and informative guide that enhances their enjoyment of this region and its wonderful environment.

Keith Broomfield

The clear waters of the Mediterranean are fascinating to explore. A tranquil spot in Kefalonia, ideal for snorkelling.

INTRODUCTION

The Mediterranean is by far the most popular holiday destination in the world, lying close to the vast population centres of Northern Europe, and holding a host of attractions including reliable hot and sunny weather in summer, ancient cities, diverse cultures and a stunning coastline.

Visiting beaches is one of the most popular activities for visitors, offering a place to relax and to enjoy a variety of water sports. One of these is snorkelling, and many visitors will have at one time or another taken a dip into the sea with snorkel, mask and fins (flippers), and become immediately enthralled by the clarity of water and the abundance of colourful and diverse marine life.

The underlying reason for such marine richness is the Mediterranean's unique geographical position, bounded by three continents – Africa, Asia and Europe – providing an interface for fauna and flora, with influences and overlap from more northern regions merging with those from southern, subtropical areas.

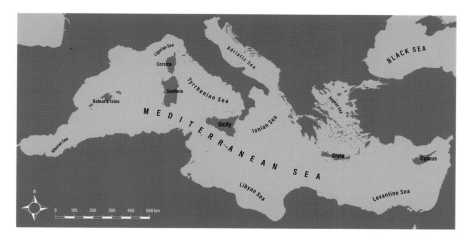

It is this melting pot which makes the Mediterranean so interesting to naturalists. The Mediterranean region is the cultural cradle of humanity and centred by a vast sea that is almost completely enclosed. The 46,000 km of coastline, which borders 22 countries, includes many islands. It is the largest and deepest of all inland seas, supporting a wealth of life, and is connected to the Atlantic by the narrow, 13 km-wide Strait of Gibraltar and to the Black Sea by the much narrower straits of the Dardanelles and Bosporus. There is also a link to the Red Sea through the artificial conduit of the Suez Canal, which has caused a range of ecological problems in more recent times, with a separate Indo-Pacific bio-fauna colonising the Mediterranean and competing with native species. These invaders from the Red Sea are termed Lessepsian migrants – and are now often seen by snorkellers, particularly in the eastern Mediterranean basin.

The Mediterranean lies at the interface between different biological zones.

The Mediterranean, which covers about 2.5 million square kilometres, can be conveniently divided into western and eastern basins that are separated by the shallow area between Sicily and Tunisia. The eastern area contains the two distinct regions of the Adriatic and the Aegean Seas, characterised by numerous islands. The western area also features islands, most notably the popular holiday destinations of the Balearics, as well as Sardinia and Corsica.

The movement of water within the Mediterranean is complex, but in simple terms, the evaporation exceeds seawater inflow, making the surface of the Mediterranean much denser and saltier than the adjacent Atlantic. The surface water sinks when cooled down, creating a conveyor-belt effect of water movement, releasing deep salty water out into the Atlantic, which is replenished by an inward flow, enabling the water in the Mediterranean to completely renew every 80 to 100 years or so. Another characteristic of its enclosed nature is the small rise and fall of the tides compared to areas bordering the open ocean. This makes shallow inshore areas a more stable environment for marine wildlife (albeit often less productive) compared to the intertidal zone in other parts of the world. Furthermore, the Mediterranean is not as rich in plankton as other oceans, which affects marine productivity, although this does ensure that the sea has a crystal clarity in many parts, making it the perfect place for snorkellers and divers to explore. Indeed, the Mediterranean is extremely biodiverse, with more than 20,000 species recorded.

The crystal clarity of many parts of the Mediterranean ensures it is the perfect place to snorkel and can be an unforgettable experience.

There is a fair degree of temperature variation in the waters of the Mediterranean, and eastern parts are generally warmer than the west. Surface water temperatures never drop below 10°C (the coldest temperatures are in the northern Adriatic in winter) and can reach over 30°C in some southern parts during summer, such as the Gulf of Sidra, off Libya. There is around a 10°C range difference between winter lows and summer highs. The higher temperature in the eastern half causes increased water evaporation and a higher salinity compared to the west.

The enclosed nature of the sea here has resulted in the evolution of a distinct ecosystem and a fascinating array of fish and other marine life. With the Mediterranean being a popular tourist destination, and the warm water being ideal for snorkelling from April through to November, it offers the perfect opportunity for people to see at first hand the array of life found in these shallow coastal waters. This marine biodiversity, especially in rocky areas, is wonderful to see, and snorkellers have long been entranced by the Mediterranean's underwater world. For many, appreciating this beauty is enough, but for others and with curiosity aroused, questions will linger in the mind as to the identification of the various creatures witnessed during a typical snorkel. The purpose of this book is to fill that void: an easy-to-use identification guide, focusing on the species most likely to be encountered, and based on the experiences of the author over four decades of snorkelling in various parts of the Mediterranean.

There will always be plenty see – even in the shallowest areas.

The nature of tourism and geopolitics means that the northern shore of the Mediterranean, along with the islands, stretching from Gibraltar in the west to Turkey and Cyprus in the east are the most accessible and visited parts of the region, and so the book focuses mostly on these areas. Tunisia and Morocco do, however, offer two notable exceptions for ease of travel on the southern coast.

Knowledge and understanding bring a greater enjoyment of the natural world, and this book is principally aimed at the casual Mediterranean holiday snorkeller who is interested in identifying the species most likely to be encountered. This is not a comprehensive and detailed guide for dedicated naturalists covering all the marine inhabitants of Mediterranean inshore waters, but instead an introduction for those with more than a passing interest in nature, offering an insight into and overview of the breadth of life to be found here. Hopefully, the book will enthuse some readers to become even more passionate about the Mediterranean, and for that audience there are other more comprehensive identification guides available.

The Mediterranean is one of the world's top tourism areas. A typical Mediterranean sandy beach.

It is also hoped that this book will play its own small part in ensuring the conservation and protection of this highly vulnerable and threatened region. The threats are many – pollution, overfishing, climate change and the introduction of invasive non-native species to name but a few. Hundreds of millions of tonnes of sewage are dumped into the Mediterranean each year and global warming could potentially prove disastrous for precious and biodiverse seagrass meadows, which stop growing if water temperatures rise too high. The Mediterranean is a sea where the hand of humanity is all pervasive and presents its greatest challenge. It is only through strategic and concerted action, fuelled by the pressure of 'people power' that the Mediterranean can fulfil its true potential as a rich and diverse marine environment which benefits both the ecosystem and humanity.

Main types of marine habitat

The abundance and types of species change depending on the habitat, with the greatest variety generally found over rocky areas, although sand and other soft bottoms are always worth snorkelling around as they offer their own range of specialised creatures and seaweeds.

Rocky shores

Rocks provide shelter and secure places for invertebrates to anchor and for encrusting algae and corals to grow. This means they are rich in life, and because there is little sediment in the water, the visibility is usually very clear. Many Mediterranean holiday resorts are located by sandy beaches, but it is common to find areas of rocks nearby. Even a small cluster of rocks and boulders acts like a haven when close to areas of sand and always merit investigation. Sea caves and rock overhangs and cliffs are typically rich in life and well worth exploring.

This rocky inlet is an ideal snorkelling site – sheltered and full of marine life.

Sandy shores

Soft-sediment seabeds such as sand and mud do not hold the same diversity of life as rocky areas, but nonetheless they always merit exploration by snorkellers. For instance, some sea breams prefer sandy bottoms and these are also good areas for Red Mullet, gobies and flatfish like the Wide-eyed Flounder.

Sandy seabeds do not have the same biodiversity as rocky areas.

Pebbles and shingle

Pebble beaches abound in many parts of the Mediterranean basin. Pebbles and shingle are highly dynamic, mobile environments, which make them challenging for creatures and seaweeds to gain tenure and so the variety of life is generally poor. In the surge zone, where the sea laps the beach, such areas are worth exploring for blennies and gobies, attracted by the rolling pebbles which release detritus into the water for these fish to feed upon.

Shingle and pebble bottoms are often the haunt of a variety of fish, including gobies such as this Incognito Goby.

Harbour walls, piers and jetties, and other artificial structures

Always be wary and use your common sense when considering snorkelling within the confines of a harbour or marina – usually this is not permitted due to the danger of being struck by boats, plus there are health hazards from sewage pollution. However, the outer walls on the seaward side can be good places to snorkel, offering the same variety of species as rocky habitats and being rich in encrusting algae.

Never venture near to a harbour entrance and be aware of boat traffic and lost hooks and lines from anglers: always respect anglers and steer well clear of their fishing activities when snorkelling. Although not often encountered in shallow waters, boat wrecks are excellent places to seek marine life as they act like underwater reefs.

The outer walls of harbour breakwaters can be productive places for snorkelling.

Wrecks are great places for snorkelling.

Open water

Open-water habitat is a good place to spot surface- and midwater-swimming fish (pelagic species) such as Garfish, Picarel and Barracuda. A tree log or other floating debris is often worth investigating as there could be fish sheltering beneath, such as young Amberjack. However, it is not advisable to venture too far out when seeking these species, always keep the shore within reach of an easy swim.

A shoal of Damselfish in open water.

Coastal lagoons

Shallow coastal lagoons with their mud or soft-sediment bottoms are often great places to go snorkelling because the still and sheltered environment offers a warm and protected place for a variety of life such as molluscs, seahorses and gobies. The visibility in coastal lagoons is often poor, however, due to sediment in the water. While lagoons are often brackish, some are the exact opposite and hypersaline due to intense evaporation.

Seagrass meadows

Underwater meadows of Neptune Grass (*Posidonia oceanica*), or seagrass as this species is colloquially known, are wonderful to explore. At first glance the beds of flat ribbon-like leaves appear uninteresting and devoid of life but look closely and all kinds of creatures are found hidden within, including seahorses, pipefish and Annular Sea Bream. Seagrass meadows form important nursery areas for fish and shellfish. Seagrass leaves are typically coated in small algae, bryozoans and hydroids, further adding to the rich biodiversity.

A seagrass meadow.

Snorkel equipment

A wonderful facet of snorkelling in the Mediterranean is that all that is required is a snorkel and mask and reasonable confidence in the water. Fins, or flippers, are desirable but not essential, and for the pretty affordable outlay of purchasing a snorkel and mask from a beachfront holiday kiosk, it is possible to gain an insightful glimpse into the underwater world of the Mediterranean.

However, good equipment enhances the enjoyment to a much higher degree and, if possible, it is worth investing in such gear. A comfortable facemask with a good seal is especially important – there are few things worse than a mask that continually fills with water, mists-up or is uncomfortable to wear. Simplicity is key, and the best option is a mask that has a double-flanged seal on a flexible silicone/ plastic skirt (as illustrated in the photograph), and which covers the eyes and nose. Choose a low-volume mask that sits close to the face to reduce water drag and enhance visibility. Also opt for a mask with a separate nosepiece to enable nostril pinching: this will help clear the ears from water pressure issues when diving to depths of more than a few metres (see more below). A good-quality, easily adjustable strap is desirable.

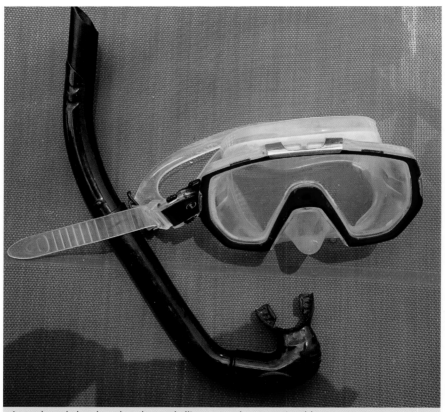

A good snorkel and mask make snorkelling so much more enjoyable.

It is best to keep the snorkel simple, too. All that is needed is basic tube with a comfortable mouthpiece. The mouthpiece can often be swivelled to ensure a more comfortable fit. Many snorkels have a valve for closure at the top of the tube when diving under, or for blowing out water by the mouthpiece, which work well. Full facemasks are also available that cover face and mouth, with an inbuilt snorkel. For some, this might be the best option, especially for those ill at ease in the water, although if possible, I would always recommend the simple separate facemask and snorkel format as highlighted above.

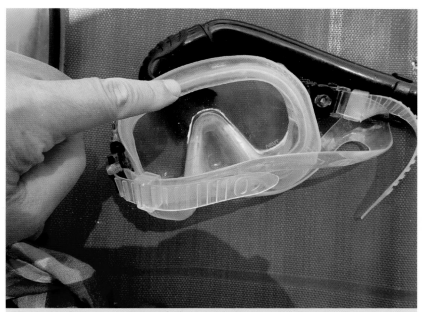

The double flange on the inside of a facemask helps prevent leaks.

The author with a valve-type snorkel.

A pair of fins transforms the underwater experience by providing much greater propulsion and manoeuvrability. There are two types available: full foot fitting like a shoe, or partial foot fitting with the heel held by an adjustable strap. Both are good and depend on individual preference. If opting for the full-shoe type, it is important to ensure that there is a tight and comfortable fit and no danger of the flipper or fin falling off. I also recommend taking beach shoes for approaching a snorkelling site, as this enables getting close to the water while offering protection to the feet over rocky areas.

Essential gear – facemask, snorkel, fins and beach shoes.

The Mediterranean is usually warm enough to snorkel in normal swimwear, but in the early part of the season – from March until early June – the water can feel cold and so wearing a neoprene bodysuit (or even full wetsuit) is worth thinking about. Issues to consider are the amount of space it might take up in holiday luggage and the buoyancy that a protective bodysuit offers. The advantage of buoyancy is that it aids safety, but the disadvantage is that this makes it more difficult to dive under the water for those inclined to do so. A weight belt overcomes this inherent buoyancy, but in most cases is impractical to carry when travelling on holiday due to airline luggage weight restrictions.

Recently, I have taken to wearing a knee-length and elbow-length bodysuit every time I visit the Mediterranean – and have found it to be a gamechanger. I can snorkel for longer because cold is not an issue, and the suit is thin and light enough to fold easily away in a suitcase. A further advantage is the protection offered from sunburn on the back. The buoyancy can be troublesome, especially when diving under in shallow water, but is less of an issue on dives deeper than 3 m.

Snorkel techniques

For those who have never snorkelled before, it is beneficial to become gradually accustomed to the marine environment by keeping to shallow water.

It is difficult to walk when wearing fins, so sit in shallow water by the shore edge and put them on there. If not wearing fins, beach shoes are a good idea, especially since Mediterranean shores are often rocky and can be razor-sharp. Beach shoes also mean one can walk comfortably down to the water's edge, before slipping into snorkel gear. If wearing full-foot fins, bend back the rubber heel of the fin so that each foot glides in smoothly and then snap the heel back up. Wetting each fin beforehand greatly aids this process by providing lubrication.

Wash the mask out in the sea and smear the inner glass surface with saliva, and rinse again. This might seem unsavoury but it helps prevent the mask misting. Another effective technique to prevent misting is to smear a thin layer of tooth-paste or washing-up liquid over the inner glass, and then wash-off (this can be done in your holiday accommodation). Put the mask on and be careful to ensure no hair gets trapped in the seal as this will provide a conduit for water leakage. Moustaches and beards can cause water leakage problems – there are environ-mentally friendly balms available that can be smeared over a moustache to resolve this issue.

Once kitted-up, turn the body and lie flat in the water face down and slowly pull away from the shore edge by using both fins and hands. Once floating free, steady the breathing and relax. If new to snorkelling, then just do so in very shallow water less than a metre deep to build up confidence – there will still be plenty to see! The key is always to stay within your comfort zone.

Two skills are required to see marine life. The first to snorkel quietly and efficiently, the second to observe and be part of the environment. When moving the fins, try not to break the water surface, as the splashing will disturb fish. Keep the body at a slight angle, with the legs sloping marginally downwards to prevent the fins splashing the water surface. Try not to bend the knees too much when moving – easier said than done, but it will provide more power. Keep the leg movements slow and purposeful, rather than fast and furious.

Diving under is not essential as so much sea life can be seen from the surface, but it does offer a different dimension for those with the confidence to do so. In order not to scare marine life, a dive needs to be executed with the minimum disturbance in one smooth movement. In order to achieve this, take two or three deep breaths, plunge the head and upper body downwards and at the same time

The most productive way to see marine life is to snorkel with gentle movements.

smoothly throw the legs into the air, keeping them straight and the feet together. Pull sharply with cupped hands to provide momentum and the body will slide under in a jack-knife dive. The head, body and legs need to act as a straight, almost vertical, arrow to execute a perfect dive. Only start kicking the fins once they are fully under the water as that will prevent surface disturbance. On resurfacing, a sharp blow into the snorkel should blast out the water in one go, making it possible to breathe again. This is a technique that requires some mastering, so practice in shallow water by dipping the head and snorkel underwater and then blowing the water out. If using a snorkel with valves at the top-end and by the mouth, you may not need to blow the water out with a sharp puff. The key is to practice under safe conditions to gauge how the snorkel works best for you.

When diving under more than a few metres, ear discomfort will be experienced. This is caused by water pressure on the external part of the eardrum being higher than that in the middle ear. To rectify this and equalise the pressure on each side of the ear drum, pinch your nostrils through the nose cover of the mask and gently blow. This should open the Eustachian tubes, small passageways that connect the throat to the middle ear and which help balance pressure. It will usually equalise the pressure and ease the discomfort, but never blow too hard and force the issue as this could cause damage. If the ears do not clear or 'pop' easily, immediately surface. Never dive under the water if suffering from a cold or sinus problems.

It is not unusual when snorkelling, especially by rocks, to feel the sea temperature suddenly drop and for the visibility to become blurry. This is caused by freshwater seeping out from a fissure in the rock; because it is less dense, this water creates a surface layer above the saltier water below. It can be annoying, but usually by swimming on it is possible to emerge from the area quickly and enter back into normal sea conditions. However, it is worth remembering that this is a phenomenon that is most frequently encountered if snorkelling near the mouth of a river, stream or other water outflow.

Always snorkel within your comfort zone and never take risks.

Safety

Safety is paramount when snorkelling. Ideally, a snorkeller should consider partaking in a training programme with a dive club or school, but because snorkelling is often a casual holiday pastime, this is understandably not always possible. Much of safety revolves around employing a common-sense approach.

Some key measures to adopt are:

- Ideally, snorkel with a companion, or have someone on the shore aware of where you are, or watching you.
- Beware of rip currents, they occur in the Mediterranean and can be dangerous.
- Do not enter the water if the sea is rough.

- Beware of sharp rocks. Many shoreside rocks in the Mediterranean are razor-sharp and it is easy to brush against them accidentally, especially if there is a surge in the sea.
- Always stay within easy reach of the shore – the most interesting marine life is invariably in shallow water, so stick to these areas.
- Never enter the sea when there is thunder and lightning.
- Do not snorkel too long, to prevent becoming cold and tiredness setting in.
- Beware of the sun on your back – wear the highest protection sunscreen possible, or a t-shirt or other protection such as a wetsuit. When snorkelling, your back is at a 90-degree angle to the sun, meaning you can get seriously burnt in only a short time without realising it is happening.
- It is best to avoid touching any marine creature. This not only prevents disturbance, but some may be poisonous. Take particular care when standing or touching rocks to ensure there are no sea urchins about, and always give jellyfish a wide berth – they often have long tentacles that can be hard to detect. Frequently look ahead of you when snorkelling in case there are jellyfish on the surface.
- Always snorkel within your comfort zone.
- Be aware of boats – a snorkeller is very hard to see in the water. Do not snorkel in areas busy with boats, such as marinas and harbours.
- If you are alone and have left a bag on the shore with clothes and other belongings, be aware of the consequences of theft. If you have left a bag when snorkelling on a remote section of shore, always take a careful note of where it has been left. There have been several occasions when the author has had problems relocating his gear bag! Similarly, be aware of gusty winds that may blow a bag away.
- When snorkelling on rocky shores, choose your entry point with great care to ensure it is also suitable for an easy exit. Many rocky shores have steep, cliff-like

Rocky shores are great for snorkelling – but be aware of the dangers.

inclines, making coming back ashore challenging unless you have chosen a safe and easy access point. Also take a careful note of what your entry point looks like and where it is, as it is easy to forget after being in the water for a while. When engrossed in snorkelling it is possible to snorkel further along a shoreline than intended – make sure you have sufficient energy reserves to get back to a starting point – which may be a tougher swim than anticipated depending on wind and current.

- Consider using a swim safety buoy that can be towed, which prominently marks your position and can be used as floatation aid.
- Learn (and for those with you) the number to call for an emergency in the European Union – 112 – it could save your, or someone else's, life. Similarly, knowledge of resuscitation techniques is a useful skill to have.

Observing marine life

Many fish and other marine creatures are surprisingly tame and easy to approach, but others are less so. Thus, to observe marine life to the optimum level it is important to become part of the environment and create minimum disturbance.

This means snorkelling slow and easy, using gentle and casual flicks of the fins, and sometimes just the arms and hands for movement. Continually scrutinise the seabed below and just ahead, is there something that looks a little bit different or

Sea caves are often rich in marine life.

out of place? If so, it could be an unusual fish resting on the seabed or some other creature relying on camouflage for concealment. The upper margins of steep rock-faces, or rock shelves, in water less than two metres deep are often productive locations to find resting blennies. These fish are well camouflaged, so approach slowly and peer into every nook and cranny. Such places are also good for colourful anemones and sponges.

Similarly, very shallow water in the splash zone of the shore less than half a metre deep can be fertile territory for exploration. By using the hands to crawl along these shallows, there is a good chance of spotting a range of unusual creatures such as gobies, blennies and prawns.

Often, on seeing an interesting fish ahead, a good ploy is to stop swimming and let your body slowly glide towards it. In many instances this gentle approach and avoidance of movement is enough to prevent scaring the creature. Sea caves and rock overhangs are always productive places to explore for groupers, Cardinalfish and other marine life, but always ensure you are confident of exiting and be aware of the risk of hitting your head on the overhang. Caution is always the watchword, and the golden rule is to take no risks.

Fish are what will initially attract a snorkeller, but to garner full enjoyment of the underwater world, it is worth taking a close interest in the numerous invertebrates and seaweeds found in the Mediterranean. This often means scrutinising the surface of rocks and cavities methodically and carefully. Adopting such a technique will never disappoint because there is always a multitude of fascinating encrustations to be discovered, as well as diminutive invertebrates.

Photography

Underwater photography is a great way to enhance snorkelling enjoyment and aid in species identification. I use a simple, compact underwater camera that is easy to carry and delivers good photographic results. There are many suitable cameras on the market, and because there are continual new product launches, when purchasing a camera the best option is to check online reviews to see what type best suits your needs and budget.

Taking underwater photographs is challenging, and in bright sunshine the digital viewing screen is often hard to see because of the reflective glare. In such situations, the camera has often simply to be pointed in the general direction of the fish and a degree of luck is required for the autofocus and camera placement to capture the subject. It is a good idea to take a lot of photographs, which increases the chance of at least a few being good quality. The best conditions for underwater photography are often when the sky is overcast, so do not be put off by cloudy days. Using a flash can also dramatically enhance the quality of photographs, although most of the time I do not, for there is usually sufficient light in shallow water.

Often when trying to take photographs while diving under, inherent buoyancy forces the body up from the seabed. In such situations, it may be necessary to grip onto a rock with one hand and hold the camera with the other – a challenging balancing act. There is also a shallow depth zone at about two metres where it is difficult to dive down to because one cannot gain momentum and buoyancy quickly throws the body back up to the surface.

The slow, gliding approach with no fin movement as previously described is a good method for getting close to fish, as well as extending the camera out to full arm length to get it as close to the subject as possible. Avoid sharp movements of the camera as this too will scare fish and other marine creatures – the overriding mantra being to do everything slowly and smoothly.

User guide

This book is intended to be an easy-to-use introductory photographic guide for snorkellers on the most frequently encountered Mediterranean marine life. It avoids scientific jargon where possible, but it is useful to know the basic external anatomy of fish and the terminology used. The diagram below highlights some of the different terms used for body parts.

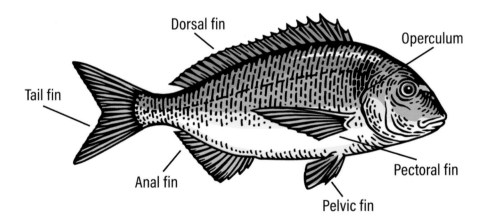

For each species, there is a short description, and information on distribution and status, and behaviour and habitat. Snorkel tips are also given on how to seek them out.

One issue with Mediterranean marine life is the nomenclature, with some species having several different English names. For example, the sea bream species with the scientific name *Sarpa salpa* can be variously known as Saupe, Salema, Cow Bream or Goldline. In such instances, the alternative names will be listed below the name deemed to be most frequently used.

A group of fish especially challenging for identifying is the gobies. There are over 60 species of goby in the Mediterranean, many of which look similar and need an expert to identify. For simplicity, only a small number of the most frequently encountered types of goby are included in this book.

As I researched this guide, it became increasingly apparent that there is a wide range of species, especially seaweeds and invertebrates such as tubeworms, sponges and sea squirts, that are very difficult to accurately identify under snorkel conditions. Indeed, even experts I consulted with on some of my photographs found it impossible to identify the species with certainty, although it was normally possible to narrow down to genus or family. So, where there are organisms that are difficult to accurately identify when under the real-life conditions of snorkelling, they have been described at the level of genus or family, rather than individual species.

One notable omission from this book is detailed accounts of sharks and rays, for the simple reason that they are so rarely encountered when snorkelling – a sad sign of their lack of abundance. However, brief descriptions have been included of a small selection of sharks and rays which a lucky snorkeller may discover.

For the order in which the species are catalogued in the book, fish are arranged with the most common families or groups likely to be encountered when snorkelling at the beginning, and with the least frequent towards the end. Invertebrates are similarly arranged.

It is often challenging to identify organisms down to species level when snorkelling.

FISH

Fish are made up of two main groups: the bony fish (Class *Osteichthyes*), and the cartilaginous fish (Class *Chondrichthyes*), which comprise sharks and rays. Bony fish are by far the largest group and are characterised by having a skeleton made of bone, gill covers and a swim bladder to aid buoyancy.

Sea breams

The sea breams are a large group of fish, commonly encountered when snorkelling, and are characterised by having laterally flattened oval bodies.

Annular Sea Bream

Diplodus annularis

Description: A small sea bream (typically 5–14 cm in length, although occasionally larger), with a vertically compressed silvery body, and yellowish pelvic and anal fins. There is usually a yellow tinge on the forehead and nape, but this is not always obvious. Black band on tail stalk. There is a black spot at the base of the pectoral fin, although this is not always easy to discern. Young fish may have indistinct, dark vertical stripes on the flanks.

Status: Common throughout the Mediterranean in suitable habitat.

Biology: Occurs most frequently over seagrass (*Posidonia*) beds, and other shallow weedy areas such as by harbour walls and in lagoons. Sometimes found by algae-encrusted boulders in very shallow water adjacent to sand. Often occurs in brackish water, especially when young. Usually appears singly, or in small, loose groups of two to six individuals. Frequently hovers in amongst or just above seagrass fronds. Feeds on small invertebrates such as worms and crustaceans.

Snorkel tips: Superficially similar to a small White Sea Bream (*Diplodus sargus*), the yellow pelvic and anal fins are diagnostic, although they are not always distinct and clear to the eye. These are energetic little fish, which dart backwards and forwards when approached, but rarely flee any distance from their preferred patch of seabed. Their quick movements can make them challenging to photograph. Seek them out in seagrass beds and other sheltered, shallow weedy areas.

Bogue

Boops boops

Description: Unlike other sea breams, this is a long, slender fish with a beige tinge to the body and a noticeable dark lateral line. There is a small dark spot at the base of the pectoral fin, but this is not always obvious. The eye is large. Typically around 14 cm in length, but can grow to 30 cm.

Status: Common throughout the region.

Biology: A surface and midwater swimmer, Bogue occur in large shoals above a variety of sea bottoms. Sometimes also in pairs or small groups. Shoals often rise to the surface at night. Will also often feed close to the seabed, including in very shallow water. Food comprises tiny crustaceans and other planktonic creatures and algae. An important commercial fish in the Mediterranean. Occasionally found in mixed shoals with Mediterranean Sand Smelt (*Atherina hepsetus*).

Snorkel tips: Bogue are easy to approach close and it is often possible to swim through shoals without undue disturbance. They are, however, challenging to photograph because of their relatively small size and fast movements. Look out for them by rocky outcrops where large gatherings often congregate.

Common Two-banded Sea Bream
Diplodus vulgaris

Description: A distinctive oval-shaped sea bream with two prominent vertical dark bands – one behind the head and the other near the base of tail. There are faint, barely discernible, horizontal golden lines along each flank. The predominant colour is silvery-grey, with a distinct blueish tinge on the head. Steep profile to forehead. Typical length 8–28 cm, although can reach 45 cm. Juvenile fish under 4 cm do not exhibit the dark bands.

Status: Common throughout the Mediterranean.

Biology: Found over algae-covered rocks from very shallow water to depths of 30 m and frequently occurs over seagrass (*Posidonia*) beds, and occasionally sand. Often encountered by rocks that are adjacent to sand. Usually found singly or in small shoals, swimming close to the bottom and among rocks to feed upon small crustaceans, worms and molluscs. Occasionally appears in midwater in small shoals.

Snorkel tips: A confiding fish that is easy to approach close. Often seen swimming with other sea bream species.

Gilthead Sea Bream

Sparus aurata

Description: A deep-bodied sea bream with a steep forehead and large black blotch on the gill cover. The end of the tail fin is often dark edged, as is the top margin of the dorsal fin. Pectoral fin is noticeably long. Background colour varies from silver to blue-grey – individuals over sandy bottoms tend to be paler. There is a golden band between the eyes on larger specimens, but this is difficult to discern when snorkelling. Can grow up to 60 cm, but most specimens are in the 15–35 cm range.

Status: Found throughout the Mediterranean.

Biology: A bottom-living fish, found over rocks, sand and mixed bottoms, and seagrass (*Posidonia*) beds. Tends to occur in

Juvenile – note that the large blotch on the rear of the gill cover is absent in young fish.

deeper areas than many other sea breams, usually from 5 m downwards. Generally seen singly, occasionally in pairs or small groups. Often occurs in brackish water. Eats crustaceans and molluscs. Fished commercially and farmed in the Mediterranean.

Snorkel tips: These are wary fish – usually shy and fleeing at speed into deeper water on the approach of a snorkeller. However, some individuals are more confiding, especially smaller fish, and will allow a close approach. Early morning seems to be the best time to encounter Gilthead Sea Bream, possibly because they move into shallower water during the hours of darkness, or more likely because the water is undisturbed by human activity at that time.

Saddled Sea Bream

Oblada melanura

Description: Has a distinctive sky-blue tint to the body when observed in certain light conditions, especially from a distance greater than two metres. However, seen close-up, the oval-shaped body appears silvery. There is a distinctive black blotch at the base of the tail, ringed with white. There are faint, horizontal stippled lines along the flanks and a prominent slightly curved, dark lateral line. Average length 10–28 cm.

Status: Common throughout the Mediterranean.

Biology: An open-water swimming sea bream, although usually staying close to rocky outcrops, or over seagrass (*Posidonia*) beds. Most frequently swims in shoals, sometimes quite large, although occasionally singly or in pairs. Shoals often found hanging motionless in midwater. Occurs from about 1 m from the surface downwards. Feeds on small invertebrates and algae.

Snorkel tips: An attractive and commonly encountered fish, the Saddled Sea Bream is shy in demeanour and hard to approach close, always intent on maintaining a reasonable distance from a snorkeller. This makes the species difficult to photograph, especially small individuals that can put on a good turn of speed when swimming in midwater.

Salema

(Saupe, Cow Bream, Goldline, Salema Porgy, Dreamfish)

Sarpa salpa

Description: Distinctive oval-shaped body, silvery with a subtle blue-grey tint and distinctive yellow lateral stripes along the flanks. Small mouth and yellow eyes. Small black spot at base of pectoral fin. Length up to 30 cm, but usually 10–22 cm.

Status: Common throughout the Mediterranean.

Biology: Found over rocky and mixed rock and soft seabeds from 1 m depth downwards. Also occurs over seagrass (*Posidonia*) beds. Almost always in shoals of same-sized individuals, sometimes swimming just above the seabed and feeding on algae on rocks and seagrass. Small, young individuals (4–6 cm) will sometimes swim close to the surface.

Snorkel tips: Salema are relatively easy to approach, especially when foraging on algae-encrusted rocks where they sometimes become engaged in a feeding frenzy that throws clouds of detritus up into the water column. On such occasions they will barely notice a snorkeller. When disturbed, they swim swiftly in a close-packed shoal, but usually soon settle back down.

Sharpsnout Sea Bream
(Sheephead or Sheepshead Sea Bream)
Diplodus puntazzo

Description: Oval silvery-grey body with distinctive long pointed snout. Body streaked with up to 13 distinctive dark vertical stripes. A black spot completely encircles the tail base. Body length typically 20–30 cm, although can grow bigger.

Status: Common in most parts of the Mediterranean.

Biology: Frequently found over rocky seabeds from 1 m downwards. Often solitary, although frequently in pairs, and occasionally in small groups, and sometimes in the company of White Sea Bream (*Diplodus sargus*). Not as frequent as White Sea Bream, although still common. Feeds on small invertebrates and seaweeds.

Snorkel tips: Similar to the White Sea Bream in size and behaviour, although the more distinctive and numerous dark vertical stripes, combined with sharp 'sheepshead' snout, usually prevents misidentification. The Sharpsnout Sea Bream is shyer than White Sea Bream, although relatively easy to approach if the snorkeller adopts slow and cautious movements. Almost always over rocks, rarely on sandy bottoms.

Striped Sea Bream

(Sand Steenbras, Marmora)

Lithognathus mormyrus

Description: Body more elongated and less oval-shaped than most other sea breams. Silvery, with up to 15 vertical stripes in an alternating pattern of darker longer streaks and shorter lighter ones. Long, marginally curved forehead, giving a long-snouted appearance. In sunny conditions, often appears very pale when over sand. Average length 20–30 cm

Status: Frequent throughout the Mediterranean.

Biology: A bottom-living fish found over sandy seabeds and other soft bottoms, as well as adjacent to seagrass (*Posidonia*) beds and in estuaries and lagoons. Feeds on small invertebrates in the surface sediment by ingesting sand particles before spitting out the debris and retaining small crustaceans in the mouth. A gregarious fish, living in loose groups on the bottom, which often comprise more than 20 individuals. Sometimes occurs singly.

Snorkel tips: A distinctive fish, this is one of the principal species to seek out in shallow, sandy areas where it is common. Striped Sea Bream are easy to approach for photographic purposes. However, because they live on the bottom, it is best to dive under to get the best results. Dive under some distance from the fish so that a careful approach can be made along the seabed.

White Sea Bream

(Sargo)

Diplodus sargus

Description: Deep, vertically compressed, silver-grey body. Distinctive black blotch at the tail stalk fin which does not quite reach the bottom edge. The end fringe of the tail fin is often edged black. Gill-cover edge features thin black margin. Around nine faint vertical stripes on the flanks (alternating between dark and lighter tones). Stripes are most pronounced in small individuals but diminish in intensity in larger fish to such an extent they can be hard to discern. Length 20–30 cm, sometimes larger.

Status: Common throughout the Mediterranean.

Biology: Found over rocky and mixed rock and sand bottoms from 1 m depth downwards. Sometimes encountered individually, but usually in small groups of three or more close to the bottom. Sometimes swims in mixed shoals with Sharpsnout Sea Bream (*Diplodus puntazzo*). Specimens over sandy seabeds are often paler than those by rocks. Feeds on crustaceans, molluscs, echinoderms and other invertebrates.

Snorkel tips: The White Sea Bream is ubiquitous and one of the sea bream species most frequently encountered by snorkellers. Easy to approach. The distinct dark vertical stripes found in smaller specimens can lead to confusion with the Sharpsnout Sea Bream.

Zebra Sea Bream

Diplodus cervinus

Description: A deep-bodied sea bream with four or five broad, dark brown or olive-green vertical bands down the silver-grey flanks. The middle bars may divide into two on the lower segment. The forehead is steep and the mouth thick-lipped. Up to 50 cm in length, but more usually around 20–25 cm.

Status: Infrequent and seldom seen by snorkellers. Most likely to be encountered in the Western Mediterranean.

Biology: Occurs over rocks from about 2 m depth, either in singly or in small groups. This species probably prefers deeper water, which is why it is not often encountered by snorkellers. Feeds on small invertebrates and algae.

Snorkel tips: The author has only encountered this fish in Spanish waters, although it does occur in other parts of the Mediterranean. The Zebra Sea Bream is very distinctive because of its broad, dark vertical bars. Striped Sea Bream (*Lithognathus mormyrus*) and Sharpsnout Sea Bream (*Diplodus puntazzo*) also have vertical stripes, but these are much thinner than those of Zebra Sea Bream. A shy fish, which generally does not tolerate a close approach.

Wrasse

Wrasse are charismatic fish with a fascinating biology – the males often building nests and guarding the eggs, and some species changing sex during their lifecycles. They are one of the most frequently encountered groups of fish when snorkelling, with several species being common over shallow, rocky seabeds.

Axillary Wrasse

Symphodus mediterraneus

Description: A small wrasse, growing up to 15 cm. There is a conspicuous black spot on the upper part of the base of the tail fin, and a much smaller spot at the base of the pectoral fin, which is not always obvious. Body colour varies greatly, but typically from greenish brown to dark brown. In the breeding season the male is reddish-brown with purple undertones, light blue spots and a yellowish head.

Status: Occurs throughout the Mediterranean although not as common as some of the other wrasses in the region.

Biology: Usually found singly from depths of 2 m and greater. Adults are encountered over rocky bottoms and seagrass beds. Feeds on small molluscs, crustaceans and other invertebrates. The male builds a nest among seaweed and guards the eggs once laid.

Snorkel tips: The Axillary Wrasse is a shy species and is difficult to approach close and will generally try to maintain a reasonable distance by the fast sculling of the pectoral fins. The best chance of getting close is when the male is guarding his nest. The large spot on the top part of the tail base is the best identifier, as is the yellowish head in the male when in breeding condition.

Black-tailed Wrasse

Centrolabrus melanocercus

Description: A small, oval-shaped wrasse (typically 8 cm in length) with a beige to pink-brown body and a black tail. In males, the edge of the tail and rear of dorsal fin is often tinged with blue.

Status: Found throughout the Mediterranean but never abundant.

Biology: Usually found singly 2 m down and deeper, especially over rocks and in areas where *Posidonia* seagrass meadows occur. This wrasse is a 'cleaner fish' and is often seen swimming towards different wrasse species and other fish such as bream to pick parasites off their skin.

Snorkel tips: The Black-tailed Wrasse is not often spotted when snorkelling. However, this is a relatively confiding fish and it is usually possible to approach quite close. It is fascinating to watch Black-tailed Wrasse swim up to other fish species, which happily hang in the water to be cleaned of parasites.

Brown Wrasse

Labrus merula

Description: Oval body with a small, fleshy mouth. Body colour varies from olive to brown, with the males in breeding condition also having small blue spots. Dorsal and tail fins are fringed with blue. Size up to 45 cm, more typically 20–25 cm.

Status: Found throughout the Mediterranean, although only encountered infrequently by snorkellers.

Biology: Lives among rocks and seagrass meadows. Seems to especially favour areas with large boulders, which provide caves and recesses in which the fish can shelter. Either solitary or in small groups. Found in depths of 3 m and below (occasionally shallower). Males build a nest and guard the eggs. Feeds on molluscs, crustaceans and worms.

Snorkel tips: The Brown Wrasse lives in slightly deeper water than many other wrasses and is a fish that one will normally have to dive down to find. Despite being widespread, it is not often seen. Look out for the fish when on the surface, before diving under. It is a shy species and will usually retreat under a rock recess if approached. Has a similar shape to the Green Wrasse (*Labrus viridis*), with which it is easily confused.

Five-spotted Wrasse

Symphodus roissali

Description: A small wrasse (12–15 cm) with chequered patterning on the body. Sometimes the patterning appears as darker horizontal bands. Background body colour varies from fawn to greenish. There is usually a dark mark between the eye and the top of the mouth. There is an obvious black spot on the central part of the tail stalk. Breeding males show some colour, including light blue, on the lower half of the head.

Status: Common throughout the region.

Biology: Often found in very shallow water over rock, especially in areas with much algal growth. Also found in seagrass (*Posidonia*) beds. Breeds in early summer and the territorial male builds a nest from seaweed, including sea lettuce if available. Feeds on molluscs, crustaceans and other small invertebrates.

Snorkel tips: The Five-spotted Wrasse is a lively fish, often zipping around rocks in shallow water in the surge zone, using its pectoral fins for manoeuvring. They are easy to approach close but tend to always be on the move.

Green Wrasse

Labrus viridis

Description: Has a longer body shape than most other Mediterranean wrasses. The head, from the tip of snout to the rear of the gill cover, is longer than the greatest depth of the body. The background body colour varies greatly from olive-green to reddish-orange or shades of brown, but always with lighter speckles. Length up to 45 cm, but specimens encountered by snorkellers are usually much smaller.

Status: Found in most parts of the Mediterranean, but uncommon.

Biology: Usually found singly, the Green Wrasse occurs in depths from 3 m and deeper over rocks and mixed bottoms. The young are often found in seagrass (*Posidonia*) beds. The largest specimens are encountered in deeper water. Can swim quite fast if the inclination takes it. The male builds a nest and guards the eggs. Feeds mainly on invertebrates such as small crustaceans.

Snorkel tips: The Green Wrasse is not often seen when snorkelling and the wide variation in body colour can be confusing, as it looks similar to the Brown Wrasse (*Labrus merula*). Key indicators are the elongated body and head, and lighter speckled pattern on the flanks. Most often seen swimming just above the seabed. This is a shy fish and will not tolerate a close approach.

Grey Wrasse
Symphodus cinereus

Description: A small wrasse, typically 8 cm in length, with a pale, limey-green to grey oval-shaped body, which shows light mottling. Often there are longitudinal dark bands along the flanks. The key distinguishing feature is a black spot on the bottom part of the tail stalk, which is usually obvious, although occasionally less so.

Status: Found throughout coastal areas of the Mediterranean, but not as common as some other wrasses.

Biology: Most frequently occurs in sheltered, soft-bottomed areas, including creeks, inlets, estuaries and lagoons where there is plenty of algal growth, including seagrass (*Posidonia*) beds. Sometimes also over rocky seabeds in shallow water. Feeds on a range of small crustaceans, molluscs and other invertebrates. The eggs are laid in a nest made by the male from detritus that has been glued together. The male guards the eggs.

Snorkel tips: The Grey Wrasse is not often recorded when snorkelling, probably because its unremarkable appearance helps it go largely unnoticed, and because of its similarity to a pale version of the much commoner Peacock Wrasse (*Symphodus tinca*). It is a shy species in the main, although it is possible to get reasonably close with a careful approach.

Ocellated Wrasse

Symphodus ocellatus

Description: Colour varies greatly, but usually a light green body, with pale blue spots and darker horizontal bands. There is a dark spot on the gill cover and another at the central base of the tail. The males are striking when breeding, exhibiting vibrant colours and the gill spot is edged with blue and red. Grows up to 12 cm long.

Status: Occurs widely throughout the Mediterranean, but never abundant.

Biology: In early summer, the male builds a nest out of algae and other detritus, which he patrols to attract a female. Once the eggs are laid, the male will guard the nest zealously. Most frequently found over algae-covered rocks in shallow water from 2 m and deeper. Also found by seagrass (*Posidonia*) meadows.

Snorkel tips: Although widespread, Ocellated Wrasse are only occasionally encountered when snorkelling and are not nearly as common as some of the other wrasses. When the male is guarding the nest, or trying to attract a female, he is bold and not scared of a snorkeller. Indeed, his attachment to the nest is so strong that he will never venture far from it, and often swims around excitedly, rising up and down and twirling in the water.

Ornate Wrasse

(Turkish Wrasse, Peacock Wrasse)

Thalassoma pavo

Description: One of the most colourful – and variable – fish in the Mediterranean, undergoing different colour changes as it grows. The body and head are slenderer and more elongated compared to many other wrasses. Body colour is an eclectic mix of blue, yellow, orange and green, with vertical linear markings. The female has a black spot on the back beneath the central dorsal fin. The male is especially striking and has a noticeably forked tail. Grows up to 20 cm.

Status: Commonly found throughout the Mediterranean and has spread to north-western parts where it was previously scarce.

Biology: Usually a solitary fish (especially males), but smaller individuals sometimes occur in loose groups. Lives in shallow water over rocky ground. Ornate Wrasse start life as females, before later transforming into males. Breeding males will guard a small harem of females. Feeds on small crustaceans, molluscs and other invertebrates. At night, the fish bury themselves laterally in sandy bottoms. Lives in shallow water from 1 m in depth. Often use their pectoral fins to help propel themselves.

Snorkel tips: Ornate Wrasse are a joy to watch and are as colourful as any tropical coral-reef fish. Although it can be approached relatively close, this is an active species, constantly on the move and swirling about, which makes photography challenging.

Peacock Wrasse

(East Atlantic Peacock Wrasse)

Symphodus tinca

Description: Typical oval and with a laterally compressed body shape, as in many wrasses. Background colour olive-green with three dark longitudinal bands on the upper half of the body, the topmost one running just under the dorsal fin, and the lowermost passing across the eye to the snout. Colour varies greatly, however, with the males brighter than the females, and some individuals appearing very pale. The head is quite long. There is a distinctive dark spot at the centre part of the tail stalk, which is one of the best aids for identification. Usually 10–18 cm in length but can grow bigger.

Status: Very common throughout the Mediterranean.

Biology: Found mainly over rocky bottoms from 1 m and below. Sometimes occurs by seagrass (*Posidonia*) meadows. Stays close to the bottom and often sculls with its pectoral fins as a means of propulsion. Feeds on a range of small marine invertebrates, often by sucking up sediment and spitting out the detritus that cannot be digested. The males make a nest out of algae and guard the eggs. Occurs either singly or in small groups.

Snorkel tips: This is the wrasse species most likely to be encountered when snorkelling over rocky or mixed seabeds. A confiding fish, it is usually possible to approach reasonably close. This wrasse often presses itself close to the underside of rocks and remains motionless if a snorkeller gets too near. The Peacock Wrasse could be confused with the Grey Wrasse (*Symphodus cinereus*), which is lighter in colour and has a dark spot on the lower half of the tail stalk, rather than in the middle.

Pointed-snout Wrasse

(Long-snout Wrasse, Long-nosed Wrasse)

Symphodus rostratus

Description: A small wrasse (usually around 10 cm long), with a noticeably long snout. The forehead is concave shaped and the body profile has a humpback appearance. Greenish-brown or reddish-brown in colour, with darker mottling and sometimes two darker longitudinal lines along the flanks.

Status: Occurs in most parts of the Mediterranean, but never in abundance.

Biology: Typically occurs in seagrass (*Posidonia*) meadows and rocky areas. It is not a strong swimmer, so prefers relatively sheltered areas. Often propels itself by quick movements of the pectoral fins. Eggs are laid in a nest made by the male. At night it maintains an upright position among seagrass fronds to rest.

Snorkel tips: The Pointed-snout Wrasse is not as frequently encountered as some other wrasse species probably because it prefers seagrass beds and tends to occur in deeper water. It often swims at a slight angle with the head pointed downwards. This is a shy wrasse that will often play a frustrating game of 'hide-and-seek' with a snorkeller, making it a difficult species to photograph.

Rainbow Wrasse

Coris julis

Description: An elongated wrasse with body colour and pattern varying depending on age and sex. The most frequently seen forms are the initial-phase young males and females, which have a white stripe running along the side of the body, which separates the darker upperparts from the paler belly. The secondary phase males (which are older females that have turned into males) feature a stunning orange jagged band running along the middle of the body, which divides azure upperparts and white underparts. The eye is red with a dark centre. Grows up to 20 cm in length.

Status: Occurs commonly throughout the Mediterranean.

Biology: Found over rocky and mixed seabeds, as well as over seagrass (*Posidonia*) beds from 1 m downwards. Rainbow Wrasse are active fish, often moving rapidly over the bottom and tumbling around rocks. The two-tone appearance of the smaller individuals is distinctive. Feeds on a range of small invertebrates. Usually occurs singly, but also sometimes in loose groups of three or four fish in proximity to one another.

A mature male.

Snorkel tips: Rainbow Wrasse are commonly met by snorkellers, although their quick movements often make them difficult to photograph. Most individuals encountered are initial-phase fish, with the colourful, larger second-phase males not nearly as frequent.

Corkwing Wrasse
Symphodus melops

Description: Black spot on the tail stalk and a deep body. Generally has a dark green to brown mottled colour, but this varies greatly depending on sex and breeding condition. Dark, comma-shaped patch behind the eye. Three, sometimes indistinct, longitudinal bars along the body. Males often have reticulated blue-and-red striped patterning on the head below the eye and on the gill cover. Length: 10–20 cm.

Status: Relatively infrequent, but has been recorded in the Western Mediterranean and northern parts of the Adriatic.

Biology: Feeds on molluscs and small crustaceans. Builds a ball-shaped nest out of algae, which the male defends. Often found in small groups.

Snorkel tips: The corkwing wrasse is more of a northern species, preferring Atlantic waters where there are thick kelp beds. The author does not recall ever encountering this species in the Mediterranean (although it undoubtedly occurs here), and given its preference for Atlantic waters it is probably most likely to be encountered in western parts of the basin. A relatively shy wrasse and difficult to approach close.

Blennies

Blennies are attractive, extremely elongated fish that live on the seabed and on sloping or vertical rockfaces. Some species are small and require careful and methodical searching.

Adriatic Blenny

Microlipophrys adriaticus

Description: A diminutive blenny reaching up to 5 cm in length. Body has alternate dark and light barring on the upper half and a pale underside. Similar in shape to Caneva's Blenny (*Microlipophrys canevae*).

Status: Occurs mainly in the Adriatic and parts of the Aegean Sea.

Biology: Prefers sheltered areas of steep rock where it lives in extremely shallow water. Feeds on small invertebrates, detritus and algae.

Snorkel tips: The restricted range, small body size and propensity to live just below the water surface ensure that this attractive blenny is often overlooked by snorkellers. It is tame by nature and easy to approach close.

Caneva's Blenny

Microlipophrys canevae

Description: A small blenny (up to 7 cm length) with a steep forehead. Most usually has a brown-green background colour with tiny spots that are arranged longitudinally. Breeding males have golden cheeks with the rest of the head blackish.

Status: Widespread but not frequently encountered.

Biology: Often occurs on steep rock walls just beneath the surface in areas that frequently become exposed to the air due to the churning action of waves. Males occupy small holes during the breeding season.

Snorkel tips: A difficult blenny to spot because it often lives right by the splash zone on steep rock outcrops. Searching these areas will often bring dividends and will reveal other species that dwell in such places.

Mystery Blenny

(Diablo Blenny)

Parablennius incognitus

Description: The two eye tentacles can be very long, especially in males, and are fringed at the back with four or five smaller frills. However, in females the tentacles are much shorter and are not so obvious. Ground colour variable, usually brownish with darker bars, the upper ones disjointed from the lower ones. Up to 7 cm in length.

Status: Present throughout Mediterranean, although probably scarcer in eastern parts. Where it does occur it is never abundant, although the small size of this fish means it is probably often overlooked.

Biology: Males can sometimes be observed inside a hole or crevice in a rockface or incline. Females lay their eggs in these shelters, which are guarded by the male. Also occurs over rocky or mixed rock and sand seabed habitats. Found in very shallow depths (0.5–2 m).

Snorkel tips: It is worth searching submerged rockfaces, shelves and slopes just below the water surface for the Mystery Blenny. The male will often rest by a hole or crevice. If disturbed he will retreat into the hole, but will often poke his head out with curiosity to observe the snorkeller. The long tentacles of the male may cause identification confusion with the Tentacled Blenny (*Parablennius tentacularis*), which is not featured in this book. Tentacled Blenny tends to be in deeper water and is larger than Mystery Blenny, and less likely to be found in small rock holes.

Montagu's Blenny

Coryphoblennius galerita

Description: Colour varies depending on habitat background, but is typically green-brown to grey, with darker bars running vertically down the flanks. There is a characteristic single fringed crest on the top of the head, although this is not always obvious. Up to 8.5 cm in length.

Status: Occurs throughout the Mediterranean.

Biology: Lives on rocks in very shallow water by the splash zone and can even survive out of water, left temporarily exposed when waves recede. Feeds on small invertebrates and algae.

Snorkel tips: Although common, Montagu's Blenny is infrequently encountered by snorkellers because of its habit of living on rocks close to the surface. Careful searching of such areas is likely to bring reward.

Peacock Blenny

Salaria pavo

Description: Typical laterally compressed blenny shape, with a bulbous head and distinctive dark spot edged with blue behind the eye. Males have a prominent head crest. Body colour ranges from beige to yellow-green, and sometimes dark green. There are dark vertical bands and thin blue lines along the body. Up to 12 cm in length.

Status: Found throughout the Mediterranean in its preferred habitats.

Biology: The Peacock Blenny favours sheltered locations such as shallow lagoons and enclosed bays, and is frequently found in brackish water. It is common, for example, in the Bay of Kotor in Montenegro. Feeds on a range of invertebrates, including amphipods, as well as algae. Found in shallow water on soft bottoms but also inhabits rocky areas by the shore, particularly by harbour walls and similar underwater constructions.

Snorkel tips: A confiding fish that is normally easy to approach close. The level of tolerance, however, can vary and some individuals will flee with a burst of speed into a rock hole or other crevice if they become spooked by the near presence of a snorkeller.

Ringneck Blenny

Parablennius pilicornis

Description: The body patterning of the Ringneck Blenny varies considerably. Some specimens, especially smaller individuals, are pale with a dark longitudinal stripe along the body. More typically the body is greenish or brown, with darker vertical bars. Occasionally yellow specimens occur. There is a small, frilled tentacle above each eye. Length to 10 cm.

Status: Found mainly in the Western Mediterranean, where it can be common in suitable habitat. Very frequent in Spanish waters.

Biology: Occurs in shallow rocky areas, especially on rock shelves and inclines just below the surface. Often found by harbour walls and breakwaters. Feeds mainly on algae and detritus. The males are often territorial.

Snorkel tips: The Ringneck Blenny is an easy fish to approach and is happy to tolerate the close presence of a snorkeller. The variable patterning can lead to confusion with other species, most notably the Long-striped Blenny (*P. rouxi*), which is not often encountered when snorkelling.

Rusty Blenny

(Red-speckled Blenny)

Parablennius sanguinolentus

Description: While this species has the typical elongated shape of a blenny, the abdomen is rounded and full, giving a distinct 'pot-bellied' appearance. Background colour varies to match the seabed; those in shallow water on rocky shores are typically pale beige, whilst those in darker areas with more seaweed are greenish. The body is speckled with small spots. There is a minute tentacle above each eye. Up to 18 cm in length.

Status: Very common throughout.

Biology: Found in shallow water over pebble and rocky seabeds, often seen resting on rock shelves. Feeds mainly on algae. Specimens living on pebbles in the surf zone are frequently pale in colour.

Snorkel tips: The Rusty Blenny is often encountered by snorkellers, and provided care and slow movements are adopted, is easy to approach close, some individuals more so than others. If disturbed, will usually only swim a short distance before coming to rest again. Larger specimens are generally shyer than smaller individuals.

Sphinx Blenny
Aidablennius sphynx

Description: A beautiful, intricately patterned blenny (4–8 cm). Beige and fawn are the dominant background colours in the males, with darker brown vertical bars and subtle azure lines and scribblings, with hints of pink. The female is similar, but more muted and greener in tone. The two tiny, thin tentacles above the eyes have a frond-like appearance.

Status: Common throughout the Mediterranean in rocky areas, including by harbour walls.

Biology: Lives in tiny holes in rocks, especially vertical faces or gently sloping inclines, often in water less than 1 m deep. Typically, just the head protrudes from its rock-hole home. In spring and summer, females lay their eggs in the narrow holes occupied by the males. A confiding and curious fish, which may fully retreat into its hole on first approach but will quickly re-emerge.

Snorkel tips: Not often seen because it is overlooked due to its small size and most of body being concealed in rock holes. It is worthwhile carefully checking shallow rock ledges, shelves and vertical faces in the splash zone.

Long-striped Blenny

(Striped Blenny)

Parablennius rouxi

Description: Typical elongated blenny shape with a black stripe on the flanks set against a pale body. Up to 7 cm in length.

Distribution and status: Probably most frequent in western and northern parts of the Mediterranean.

Biology: The Long-striped Blenny lives in shallow water in rocky areas. Often inhabits holes in rocks.

Snorkel tips: Not often encountered by snorkellers, the Long-striped Blenny can easily be confused with the Ringneck Blenny (*Parablennius pilicornis*), which sometimes exhibits a body pattern variation that has a similar striped appearance.

Tompot Blenny

Parablennius gattorugine

Description: A large, robust blenny, with a prominent branched tentacle above each eye. Ground colour varies, typically yellow-brown, green-brown or red-brown. There are several darker bars on the flanks. Up to 25 cm in length.

Status: Occurs widely in the region, but possibly scarcer in south-eastern parts of the Mediterranean basin.

Biology: Present over a variety of rocky seabed habitats and often found on rock inclines. Feeds on crustaceans, worms and other invertebrates. The eggs are laid in a rock crevice by more than one female, and are guarded by a male. Usually found from depths of 3 m and greater.

Snorkel tips: The Tompot Blenny relies on its cryptic colouration for camouflage, which enables a reasonably close approach. However, they are generally wary fish and will flee if you get too near. They are bigger than many of the other blenny species encountered in the Mediterranean. The single frilled tentacle above each eye is one of the key identification features.

Zvonimir's Blenny
Parablennius zvonimiri

Description: A small blenny, about 6 cm long. Background colour varies, but usually grey or green; breeding males are red. There are 6 to 8 noticeable white spots extending along the top of the back. The forehead is steep.

Status: Widespread but only in low numbers wherever present.

Biology: This species likes shady areas in among rocks from 0.5 m and deeper, particularly by crevices and other dark recesses.

Snorkel tips: Zvonimir's Blenny is probably often overlooked by snorkellers because it is an inconspicuous and small fish, with good camouflage that blends in well with surrounding rocks. The best ploy is to search rock overhangs and vertical walls carefully from just below the surface. It is easy to approach close.

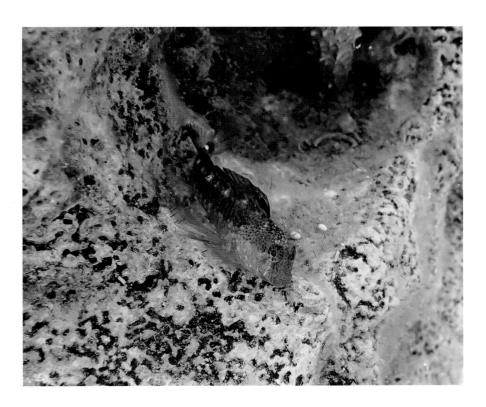

Small Triplefin Blenny
(Pygmy Black-faced Triplefin)

Tripterygion melanurus

Description: Superficially similar to the Red-black Triplefin (*Tripterygion tripter-onotus*) but much smaller, with a maximum length of 5 cm. Male and female do not differ. The head is black with pale, sometimes blue freckling, and the body is red. The forehead is not as steep as in Red-black Triplefin.

Status: Found in most parts of the Mediterranean, although never abundant.

Biology: This species likes dark, shady places, particularly the entrances of sea caves and other rocky crevices and indentations, often in shallow water from 0.5 m. Males are territorial in the breeding season. Eggs are guarded by the male.

Snorkel tips: A species to look out for in dimly lit areas, often seen clinging to the walls or ceilings of dark recesses. A confiding fish that is easy to approach close.

Red-black Triplefin

(Red Black-faced Triplefin)

Tripterygion tripteronotus

Description: A small blenny with three dorsal fins. Breeding males have a red body and elongated first dorsal fin, and a black head (males sometimes don't exhibit black heads). Females are greenish to light brown, with a slight, vertically barred appearance on the flanks. There is a short, thin tentacle above each eye, although this is hard to discern. Typical length 6–7 cm.

Status: Found throughout the Mediterranean, but most frequent in central and eastern parts of the region.

Biology: Lives in shallow depths on rocky shores, particularly on rock shelves, cliffs and inclines where it is often seen resting. Feeds on tiny invertebrates. Usually found singly, although a rock shelf may hold several individuals within close proximately of one another.

Snorkel tips: Red-black Triplefins are attractive little fish, with the bright red of the male particularly striking. They are easy to approach close and photograph. The green females can be hard to find, but careful searching of algae-encrusted rockfaces will usually bring rewards.

Top: male. Bottom: female.

Gobies

Gobies are mostly bottom-dwelling fish and represent an extremely species-rich group, with over 100 different species recorded in the north-east Atlantic, Mediterranean and Black Sea. They are generally small, living on both soft sediment and rocky bottoms, and are well camouflaged to match their seabed habitat.

Black Goby

Gobius niger

Description: A heavy-bodied goby, with breeding males being black or near black, but females and non-nesting males having a highly variable mottled pattern of darker and lighter greenish or brown tones. The most distinctive feature is its elongated, pointed first dorsal fin, which curves backwards, although this does not occur in the female. There is a dark spot on the forepart of the first and second dorsal fin, but this is not always obvious. Length: up to 18 cm, usually smaller.

Status: Common throughout the Mediterranean.

Biology: Lives on the bottom, including areas of rock, sand, mud and seagrass (*Posidonia*) beds. Occurs in shallow water down to 10 m, and often found in harbours, estuaries, lagoons and other areas of brackish water. Feeds on small invertebrates and occasionally small fish. Usually found singly, resting on the seabed close to cover of some sort. The eggs are guarded by the male.

Snorkel tips: The male when in prime breeding condition with his sooty-black head and body is unmistakable, although often the males – like the females – are more muted in tone. This goby is most likely to be spotted in sheltered, shallow-water areas, rather than exposed rocky coasts. It is normally possible to approach close if slow and cautious movements are made. Look out for the elongated first dorsal fin, as this is one of the key aids for identification. Lagoons are great places to spot this charismatic fish.

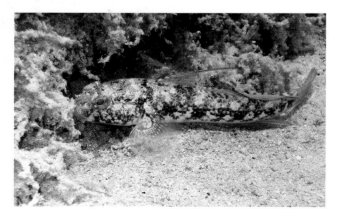

Giant Goby

Gobius cobitis

Description: A very large, heavy-bodied goby, reaching up to 27 cm. The body is a patterned, mottled grey, with indistinct dark patches on the side. The mouth is large, and the eyes look small in comparison to the size of the head.

Status: A common goby that is found in most parts of the Mediterranean.

Biology: The Giant Goby often rests on the surface of rocks and boulders, frequently near rockfaces and overhangs. Prefers sheltered parts of the coast, such as creeks, bays and inlets, and often found in brackish water. Feeds on invertebrates, small fish and algae, and behaves as an ambush predator waiting for prey to pass. The eggs are laid on the undersides of rocks and are guarded by the male.

Snorkel tips: Often encountered when snorkelling in shallow, rocky, sheltered areas. It is a shy fish, but because of its size, is easy to spot from a reasonable distance away. With a slow approach, it is usually possible to get close, but once the fish realises it has been spotted, it will quickly swim away in a rapid burst of speed, often seeking shelter in a rock crevice or hole.

Incognito Goby

Gobius incognitus

Description: This species was formerly confused with Bucchich's Goby (*Gobius bucchichi*) and was only recognised as a separate species in 2016, with the realisation that it is much more widespread than Bucchichs's Goby. A medium-sized goby up to 10 cm in length but usually slightly smaller, patterned with tiny spots that are arranged in longitudinal lines along the body. Colour varies depending on background substrate, but most commonly a sandy, pale brown, although can be darker.

Status: Found throughout the Mediterranean.

Biology: Occurs commonly resting on a variety of shallow bottoms, including rock, pebbles and mixed seabed areas of sand and rock.

Snorkel tips: Despite only recently recognised as a separate species, the Incognito Goby is probably one of the gobies most likely to be encountered when snorkelling. It is so similar to Bucchich's Goby that telling the two apart is extremely challenging under snorkel conditions, with there just being subtle differences in the patterning on the head. One nuance is that Bucchich's Goby tends to have an unmarked area on the centre of the cheek below the eye, whereas Incognito does not. The range of Bucchich's Goby seems to be confined largely to Adriatic (although it probably occurs more widely), whereas Incognito Goby is present throughout Mediterranean. Both Incognito and Bucchich's Gobies are also similar in appearance to Sarato's Goby (*Gobius fallax*), which further confuses identification!

Painted Goby

Pomatoschistus pictus

Description: A small goby up to 6 cm in length, with pale-brown background colour. The body is patterned with a series of pale saddle-shaped markings along the back, with two darker blotches below each one. Dorsal fins display black, rosy and blue horizontal streaks (not always visible). The eye is green.

Status: This species appears to have a limited distribution in the Mediterranean, occurring in northern areas such as southern France and the northern Adriatic.

Biology: Prefers sand and gravel seabeds where it will rest on the bottom. Feeds on small invertebrates. Males build a nest and entice the female to lay eggs in a courtship ritual that includes visual and acoustic signals. The male defends the next while the eggs develop.

Snorkel tips: As with many gobies, this is a difficult species to identify with certainty when snorkelling. Often the best identification technique is to take a photograph that can be examined closely later. This goby allows a close approach and can sometimes be found in loose groups resting on fine sediment areas.

Red-mouthed Goby

Gobius cruentatus

Description: A large goby (up to 15 cm in length) with red markings on the lips and cheeks. Several dark bands on the back, separated by white saddles. The head and body usually appear dark when seen from a distance.

Status: Widespread in the Mediterranean.

Biology: Found in mixed sand and rock habitats and among seagrass (*Posidonia*) meadows – sometimes as shallow as 2 m, but usually much deeper.

Snorkel tips: This striking goby with red lips is not often encountered when snorkelling because it mostly occurs in deeper water than some other gobies. Usually possible to approach quite close, provided slow and careful movements are adopted.

Rock Goby

Gobius paganellus

Description: A stoutly built goby with a large head and a dark brown, mottled body. The first dorsal fin has a pale band along the top edge. Background body colour varies greatly, typically dark brown or green, with alternating dark saddles along the back, along with mottled blotches. Length 8–12 cm.

Status: Found throughout the Mediterranean.

Biology: Occurs over algae-covered rocks, often in very shallow water. Has a liking for areas of shade, and often occurs at the entrance of small caves, rock holes and crevices. However, it can also be found over soft sediment bottoms, including shallow lagoons. Feeds on a range of small invertebrates, fish larvae and algae. The eggs are guarded by the male. Often occurs in brackish water.

Snorkel tips: Despite being relatively common, this species is often missed by snorkellers due to its effective camouflage. The Rock Goby can easily be confused with a small specimen of the Giant Goby (*Gobius cobitis*).

Slender Goby

Gobius geniporus

Description: As the name indicates, this goby has a slender body and can grow quite long – up to 16 cm. There is usually a dark patch below the eye, and the overall body colouration is greenish-brown, with a regular series of dark blotches along the side.

Status: Found throughout the Mediterranean, possibly more frequent in the western parts.

Biology: Lives inshore on sand or mud, and often by seagrass (*Posidonia*) meadows.

Snorkel tips: While probably a common species in suitable habitat, the Slender Goby is one of those fishes that goes largely unnoticed. Shallow soft-sediment bottoms with plenty of algal growth are the best places to seek this species out, which is easy to approach and photograph.

Common Goby

Pomatoschistus microps

Description: A small, pale goby (light grey to fawn in colour), with short, steep head profile, and darker bars or blotches down the sides. Often shows a darker mark by the front of the pectoral fins. The tail stalk is elongated. Length: 6–7 cm.

Status: Occurs in the Western Mediterranean.

Biology: In the Mediterranean, the Common Goby is mainly encountered in coastal, brackish lagoons, such as those found in the south of France and Spain. Also occurs in estuaries, and occasionally other sheltered coastal areas. Found more widely in Northern European waters.

Snorkel tips: Due to the Common Goby's preference for lagoons in the Western Mediterranean, it is not often seen by snorkellers. Almost always found in very shallow water, but is worth looking out for when snorkelling near brackish-water areas in France or Spain.

Grey mullets

Grey mullet are frequently encountered when snorkelling and have characteristic silvery-grey bodies. There are several species in the Mediterranean, and they can be challenging to identify under snorkel conditions because of their similarity in appearance.

Boxlip Mullet

Oedalechilus labeo

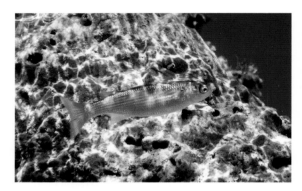

Description: A small and gregarious mullet that grows up to 25 cm but is usually much smaller. Silvery body with very faint brown horizontal lines. Mouth has large fleshy, prominent lips, the upper lip especially noticeable.

Status: Occurs commonly throughout the region.

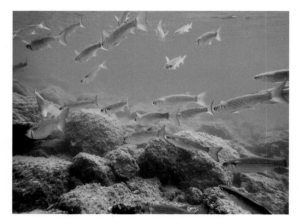

Biology: Found in shallow water over a variety of different seabed habitats. Often swims in small schools in shallow rocky areas where there is a strong surge. Feeds on detritus and algae.

Snorkel tips: The Boxlip Mullet is a lively fish often seen swimming close to the surface or swirling around in fast-swimming shoals in turbulent, shallow water close to rocks, often in very shallow water. They are confiding fish, but their fast movements can make this species difficult to photograph.

Golden Grey Mullet

Liza aurata

Description: The Golden Grey Mullet is a sleek looking fish, with a golden-yellow blotch on the gill cover. Under snorkel conditions, it can easily be confused with the similar looking Thinlip Grey Mullet (*Liza ramada*), which is not separately described in this book, and Thicklip Grey Mullet (*Chelon labrosus*). One way of differentiating from Thinlip Grey Mullet is the shape and length of the pectoral fin – the Golden Grey has long and pointed pectorals, whereas those of the Thinlip are short and rounded, with a dark spot at the base. Golden Grey Mullet may grow up to 35 cm in length, Thinlip up to 55 cm.

Status: A commonly encountered fish in the Mediterranean.

Biology: Occurs in a variety of shallow, coastal habitats, although most frequent over areas of rock. Feeds on algae, invertebrates and detritus.

Snorkel tips: Golden Grey Mullet is often encountered when snorkelling, frequently swimming in small shoals close to the surface by rocks. When snorkelling it is difficult to differentiate this species from the Thicklip and Thinlip Grey Mullets, especially since the golden-yellow patch on the gill cover is not always easily discernible.

Thicklip Grey Mullet

Chelon labrosus

Description: The Thicklip Grey Mullet is slightly chunkier in appearance compared to the Golden Grey Mullet (*Liza aurata*) and Thinlip Grey Mullet (*Liza ramada*), and the upper lip is noticeably thick – slightly greater than the diameter of the eye pupil. The flanks feature several dark, thin horizontal lines. Thicklip Grey Mullet grow up to 55 cm in length.

Status: Common throughout the Mediterranean.

Biology: The Thicklip Grey Mullet is commonly found in shallow-water areas and shows a distinct liking for harbours and marinas with sand and mud bottoms, although Golden Grey and Thinlip Grey Mullets occur in such places too. Also found in brackish water, such as lagoons, or in estuaries or streams by the coast, and will often enter rivers. Feeds on algae and small invertebrates.

Snorkel tips: Frequently encountered when snorkelling, this species is relatively easy to approach close, although its fast swimming and sudden changes in direction make these fish challenging to photograph. They often swim very close to the surface, including in open water away from the shoreline.

Pipefish and seahorses

Belonging to the family Syngnathidae, pipefish and seahorses are curious-looking fish with their external skeletons of bony plates and elongated snouts. Pipefish are slender, drawn-out fish, whilst Seahorses are compacted versions of this, with a prehensile tile and the body often held vertically with the head set at an angle.

Mediterranean Deep-snouted Pipefish

(Broadnosed Pipefish)

Syngnathus typhle rondeleti

Description: Extremely elongated, laterally compressed body that resembles a leaf of seagrass (*Posidonia*), which aids concealment. The snout is diagnostic and is long, deep and laterally flattened. Greenish or mottled brown body colour, with fainter banding and darker scribbles. Up to 30 cm in length.

Status: Found throughout the Mediterranean and probably very often overlooked because of its excellent camouflage.

Biology: Lives in shallow water from less than a metre deep in seagrass beds and over sheltered areas of mixed rock and sand bottoms. Sometimes occurs in brackish water. Often seen lying motionless on the bottom, resembling an underwater leaf. A slow and cautious swimmer. Feeds on planktonic organisms such as copepods. The male carries the eggs in a brood pouch and continues to provide parental care after hatching.

Snorkel tips: Although difficult to spot, this species is extremely easy to approach close because it is a poor swimmer and prefers to rely on its camouflaged appearance to remain unde-tected. It is worth checking out what initially appears to be a dead frond of seagrass lying on the seabed, as there is every possibility it may be a pipefish!

Greater Pipefish

(Great Pipefish)

Syngnathus acus

Description: Long, thin body with angular body rings that give an 'armour-plated' appearance. There is a conspicuous bump on the top of the head and a long tubular snout. Dark green to grey-brown body colour. Up to 40 cm in length.

Status: Found throughout the Mediterranean in suitable habitat.

Biology: Prefers to live in shallow water in seagrass beds and other weedy areas, but also sometimes found over mixed-bottom habitats. A slow swimmer that likes to hide amongst seaweed and other areas with thick algal growth. The males carry the eggs, which may be laid by several females, in a brood pouch on the underside. Feeds on a range of tiny items such as small crustaceans, fish eggs and fry.

Snorkel tips: The Greater Pipefish is the master of camouflage and is only usually ever discovered by a snorkeller through luck. The best chance of spotting this species is to carefully search areas of seagrass and locations with thick algal fronds. This is always a good habit to get into because other interesting species such as seahorses (*Hippocampus* spp.) are also found in such places.

Short-snouted Seahorse and Spiny Seahorse
Hippocampus hippocampus and *H. guttulatus*

Description: Seahorses are unmistakable fish with their upright bodies, distinctive snouts and curled, prehensile tails. There are two species in the Mediterranean. The Short-snouted Seahorse is the smaller of the two (typically 10–14 cm) with a shorter mouth, no more than a third of the length of the head. The Spiny Seahorse (body length 12–18 cm) has a snout more than a third the length of the head and often has distinctive skin filaments on its back. The body has an external skeleton of bony plates.

Status: Found throughout the Mediterranean, although infrequently seen.

Behaviour and habitat: Both species prefer sheltered environments, especially lagoons and shallow bays, and often occur over seagrass (*Posidonia*) meadows. Also found over soft bottoms, especially where seaweed detritus has accumulated, and tolerant of brackish water. The female deposits her eggs into the male's brood pouch on the abdomen where they remain until hatching after about three weeks. Seahorses feed on planktonic organisms, including tiny crustaceans.

Snorkel tips: Seahorses are probably more common than is realised, but their cryptic camouflage and skin fronds which help break up the body outline make them difficult to spot. They often remain motionless, with tail curled around a frond of seagrass or other object. It is worthwhile closely examining seagrass beds, especially in sheltered areas and lagoons. Occasionally, seahorses can be encountered swimming, or floating, just below the surface on calm days over shallow bottoms. Seahorses face many threats, including climate change, disturbance at boat anchorages, destruction of seagrass meadows and pollution.

Short-snouted Seahorse

Spiny Seahorse

Other bottom-dwelling fish

Brown Meagre
Sciaena umbra

Description: Brown-bronzed body colour with a yellowish tail. The scales on the flanks are noticeable. A heavily built fish that grows up to 40 cm in length, but more typically 20–30 cm.

Status: Occurs throughout Mediterranean but rarely seen by snorkellers.

Biology: Usually found over rocky bottoms in water that is too deep for snorkelling, but occasionally encountered in small groups in shallower water from about 4 m deep, especially steeply sloping boulder fields where there are plenty of dark spaces to hide. Feeds on small fish and invertebrates. Uses its swim-bladder to make underwater 'drumming' sounds.

Snorkel tips: The Brown Meagre is mainly active at night, although there is always a chance of encountering this attractive fish by diving down to the bottom of rocky seabeds, especially where there are dark crevices and holes.

Cardinalfish

(Mediterranean Cardinalfish)

Apogon imberbis

Description: An attractive small reddish-orange fish (10–12 cm) with a slightly rotund appearance. The large black eyes contrast starkly with the body colour.

Status: Occurs throughout the Mediterranean by rocky shores. Most frequent in central and eastern parts.

Biology: Found hovering either singly or in small groups by the entrances of sea caves or under dark rock overhangs. Occasionally in shallow water of only 2 m in depth, but usually deeper. At night, Cardinalfish leave their rocky shelters and hunt for crustaceans and small fish in more open water. After spawning, the male takes the egg mass into his mouth for protection. To provide aeration, he 'munches' his large, distended jaws to aid water movement over the eggs.

Snorkel tips: Difficult to see from the surface because of their preference for hanging almost motionless in dark, shadowy places. Best approach is to dive under and search large rock cavities and small caves and large spaces. Often quite shy and will retreat further into a cave or rock cavity if you approach close.

Dusky Grouper

Epinephelus marginatus

Description: A heavy-bodied, rotund fish with large head and mouth. Background colour ranges from a dark green-brown to chocolate brown that almost verges upon black, with paler mottling and blotches. There is lighter edging to the pectoral, anal and tail fins. Can grow over 1 m in length, but those encountered by snorkellers are usually younger specimens ranging from 10 cm to 30 cm.

Status: Occurs along most parts of the Mediterranean coast, although the population has been much reduced by overfishing.

Biology: Found mainly over rocky bottoms, especially sloping boulder fields, in caves and recesses and by loose jumbles of rocks. Juveniles can occur in shallow water (2 m deep) but larger specimens prefer deeper water. Lives singly and is territorial. Feeds on fish, squid, molluscs and crustaceans. Groupers start off life as males and change into females as they get older.

Snorkel tips: These magnificent fish are often found either sheltering in underwater caves or resting by the entrances of recesses where they can quickly retreat. Large fish are usually very shy and will not tolerate a close approach, but juveniles, while still wary, are slightly more confiding. Snorkelling early in the morning is a good time to seek these fish, as they often hunt at night, and may still be cruising over the seabed after sunrise. Due to their territoriality, they have favoured spots where snorkellers often see the same individual repeatedly, which can prove useful when engaging in underwater photography.

Goldblotch Grouper

(Golden Grouper, Striped Grouper)

Epinephelus costae

Description: Has a prominent lower jaw and in young fish there are four or five horizontal pale stripes along the body, which extend onto the upper part of the gill covers. Light brown or grey background body colour. Length up to 100 cm, although those seen by snorkellers are usually much smaller.

Status: Found throughout the Mediterranean.

Biology: Usually found over rocky seabeds and sometimes in pairs or small groups. This species is more likely to swim in open water than the Dusky Grouper (*E. marginatus*). Usually occurs from 3 m depth but tends to prefer deeper water than the Dusky Grouper and is not often seen by snorkellers.

Snorkel tips: The Goldblotch Grouper is a shy fish that will usually seek shelter under a rock or in cave on the approach of a snorkeller. However, it will often stay near the opening of its shelter, enabling reasonable views to be obtained. As this species generally avoids shallow water, the best way of seeing it is to dive under and explore likely looking areas, such as boulder fields or sea caverns.

Comber

Serranus cabrilla

Description: Resembling a small grouper, this species has a yellow-brown body, with several dark vertical stripes with one dark band running along the centre of each flank, which is mirrored with white below. A dark band may extend through the eye, and the mouth is large. Length up to 25 cm, although those encountered by snorkellers are usually much smaller.

Status: Wide distributed throughout the Mediterranean, although never abundant.

Biology: Lives on rocky bottoms from 3 m downwards and feeds on a variety of prey, including small fish and squid, worms and crustaceans. A solitary and territorial species, which is faithful to its home site.

Snorkel tips: Similar in shape to the much commoner Painted Comber (*Serranus scriba*), this species can easily be distinguished by its browner appearance and body stripes. Often described as being a curious and confiding species, allowing for a close approach – although the author has found it to be relatively shy in the main. Normally occurs in deeper water than the Painted Comber.

Painted Comber

Serranus scriba

Description: A distinctive, easily identifiable fish with bold vertical bars running down the palish flank, with yellowish tail, ventral and pectoral fins. There is a bluish blotch halfway along the lower part of the abdomen. A stripe runs through the eye, and the mouth is comparatively large. Intricate blue and red scribbling on the snout, although this is difficult to discern when snorkelling. Up to 25 cm in length, although usually about 15 cm.

Status: Common throughout the Mediterranean.

Biology: This is territorial, solitary species that is often seen hovering near the seabed over rocks, usually with a suitable hiding place nearby. Also found on seagrass (*Posidonia*) beds. Occurs over shallow bottoms from 2 m depth. The Painted Comber is at the same time both male and female (simultaneous hermaphrodite). Feeds on fish, crustaceans and other invertebrates.

Snorkel tips: Often encountered, with its zebra-like patterning quickly catching the eye. These fish often show individual personalities, with some being wary whereas others are more confiding, allowing a reasonably close approach. Their territoriality means it is usually possible to see the same individual over the same part of the seabed day after day.

Sea Bass

Dicentrarchus labrax

Description: A strong-bodied, silvery fish that can grow up to 100 cm, but those encountered by snorkellers are more usually up to 30 cm. Young fish, smaller than 20 cm, often have dark spots on the flanks which can lead to confusion with the Spotted Sea Bass (*Dicentrarchus punctatus*) which also occurs in the region, although this species usually has a distinctive dark spot on the gill cover, which is absent in the Sea Bass.

Status: Found throughout the Mediterranean, although it is declining due to overfishing.

Biology: Can be found in very shallow water, either singly, in pairs or in small shoals, swimming close to the bottom. Adults tend to be more solitary in behaviour. Young fish may sometimes congregate in larger shoals that swim in midwater. Feeds mainly on fish, crustaceans and molluscs. Often encountered in brackish waters and estuaries.

Snorkel tips: Sea Bass are not frequently met by snorkellers, although hotspot areas occur where sightings can be frequent and regular. Small Sea Bass are approachable, while adults are very shy and will usually flee at first sight of a snorkeller. Sometimes individuals will be found swimming with shoals of Grey Mullet. Young fish can occur in very shallow water, sometimes in the surf zone as they search for food items stirred up by the waves.

Spotted Sea Bass

Dicentrarchus punctatus

Description: Spotted Sea Bass can easily be confused with juvenile Sea Bass (*Dicentrarchus labrax*), which also have spots on the flanks. However, the spots on Spotted Sea Bass tend to be larger and more numerous, and there is a large dark mark on the gill cover.

Status: Probably most likely to be found in southern and eastern parts of the Mediterranean, but never as frequent as Sea Bass.

Biology: Spotted Sea Bass, especially smaller specimens, often swim in shoals in midwater over a variety of seabeds, including sand. They feed upon fish and a range of invertebrates, including molluscs and crustaceans.

Snorkel tips: When shoaling, Spotted Sea Bass are easy to approach as they seem to have confidence in one another's company. Look out for the dark spot on the gill cover to prevent confusion with Sea Bass.

Moray Eel

(Mediterranean Moray)

Muraena helena

Description: A long, sinuous and powerful body, which is beautifully marked with a 'leopard-skin' marbling of yellow and dark blue-grey. Head is long with a large, prominent sharp-toothed mouth. Up to 1.5 m in length, but more typically 60–80 cm.

Status: An uncommon fish that occurs in all regions of the Mediterranean.

Biology: A nocturnal and territorial species, the Moray Eel likes to rest during the day in rock holes, crevices and recesses on the seabed, often with just the head protruding. Feeds on fish, crabs, octopus, squid and cuttlefish. Young eels can be found in very shallow water from 1 m deep. Larger specimens prefer deeper water from about 3 m and below.

Snorkel tips: The Moray Eel is always an exciting fish to find and will frequently open its jaws and display its sharp needle-like teeth to a snorkeller that approaches close. Whilst the chances of getting bitten are extremely small, it is worth treating Moray Eels with respect, especially if the eel is cornered and has nowhere to retreat. Although not especially common, they are frequently found by snorkellers over rocky seabeds. They often have a favourite lair, and once located, it is possible to return to the same place over the following days to observe this fascinating creature. The Moray Eel is reasonably easy to approach close, although will retreat fully under cover if scared. Some individuals tend to be bolder than others.

Parrotfish

(Mediterranean Parrotfish)

Sparisoma cretense

Description: Unusually for a fish, the males are duller than the females. The male is greyish with a dark blotch behind the gill cover, while the female is red over much of body with some yellow, and a pale grey patch extending down the body from the just behind the head. The scales are large and prominent. The mouth is large-lipped. Up to 45 cm in length, more commonly 18–25 cm.

Status: Most frequently found in the Eastern Mediterranean, where it can be locally common. It may now be spreading its range in the Mediterranean due to warming sea temperatures.

Biology: Found in small groups over rocky bottoms. The Parrotfish especially prefers sloping clusters of loose rock and small boulders extending from shallow water. Feeds on encrusting algae on rocks with its rasping teeth. Uses its pectoral fins to scull its way through the water. Parrotfish always keep close to the seabed and seldom rise any distance above the bottom.

Snorkel tips: This is a charismatic species that is well-named with its bright colouration and parrot-like beak. It is usually possible to swim relatively close to a group of Parrotfish, but they are wary and will always try and maintain a reasonable distance from a snorkeller.

Red Mullet

Mullus surmuletus and *M. barbatus*

Description: These two species are similar in appearance. *M. surmuletus* – or the Striped Red Mullet – has a coloured first dorsal fin that is patterned with bands, whereas *M. barbatus* has a pale, unmarked first dorsal fin, as well as a steeper forehead. There are two prominent barbels on the chin of both species.

Status: Common throughout the Mediterranean.

Biology: *M. surmuletus* generally occurs in shallower water than does *M. barbatus* and is the species most likely to be encountered when snorkelling. Both species occur singly or in small groups, but M. barbatus appears to be more gregarious and can appear in larger aggregations. Uses the paired sensory barbels on the chin to detect food, and then catches its prey by digging into the sand.

Snorkel tips: Red mullet are easy to approach close and often become totally engrossed in their feeding, when they can be seen ingesting and then spitting out soft debris. The two red mullet species are very similar – the best indicators for separating these species being the shape of the forehead and the patterning on the first dorsal fin.

Scorpionfish

Scorpaena maderensis and *S. porcus*

Description: These two species of scorpionfish look superficially similar under snorkel conditions, and are thus included together in the one account. The colour often varies to match their background environment, but the **Madeira Scorpionfish** (*Scorpaena maderensis*) is usually brick-red, with paler mottling and a distinctive dark vertical band on the tail. The **Black Scorpionfish** (*S. porcus*) is usually brown or greenish and has a skin flap above each eye (although not always obvious). Both are stocky fish, with subtle bars, mottling and blotches on the body, and a long, spiny dorsal fin. The head, eyes and mouth are large. Length 10–20 cm.

Status: Widespread and locally common throughout the region.

Biology: Scorpionfish occur from 1 m depth and are often seen resting motionless on rocks, especially on ledges on sloping inclines next to dark recesses. They are solitary fish and feed upon a range of small fish and invertebrates.

Black Scorpionfish

Snorkel tips: Scorpionfish have highly venomous spines and should be treated with respect when snorkelling. As a rule, when placing hands upon rocks, check carefully before you do so, in case you inadvertently touch a Scorpionfish (or sea urchin). However, with a degree of caution, Scorpionfish are easy to safely approach close, as they rely upon their camouflage to remain undetected.

Madeira Scorpionfish

Atlantic Lizardfish

Synodus saurus

Description: An aptly named grey-green coloured fish, with a lizard-like head and body. Large mouth and head and an elongated, tapering, prominently scaled body. Up to 35 cm in length.

Status: Found throughout the Mediterranean, although localised in its distribution.

Biology: A nocturnal hunter that lies half-buried in soft sediment or still on rocks in readiness to ambush prey, primarily small fish. Occurs in water as shallow as 1 m, but usually deeper.

Snorkel tips: The Lizardfish is seldom seen when snorkelling, probably because of its secretive habit of resting well hidden on the seabed. At first it will rely on its camouflage to remain undetected, but when a snorkeller approaches too close, the Lizardfish will disappear in a rapid burst of speed.

Mediterranean Sand Smelt

Atherina hepsetus

Description: A small, silvery and elongated fish, usually about 10 cm in length. Greenish back and silvery, almost translucent flanks. Sand smelts in the Mediterranean are a complex group, and there is still confusion over differentiation and classification between species, including *Atherina boyeri* and *A. punctata*, which makes accurate identification when snorkelling difficult.

Status: Abundant throughout the Mediterranean.

Biology: Lives in medium-sized, sometimes large shoals in shallow water and usually close to rocks.

Snorkel tips: Sand smelts are wonderful fish to watch, with their flickering, silvery shoals having an almost hypnotic quality. They are easy to approach close, but their small size makes photography challenging.

Wide-eyed Flounder
Bothus podas

Description: The Wide-eyed Flounder is rounder in shape than many other types of flatfish. The eyes are situated on the left side of the head and are widely spaced apart. Body colour varies depending on the background environment; it is most usually a pale grey that blends in well with light sandy bottoms, but can be much darker. There is usually a mottled pattern of lighter spots on the back. Length up to 20 cm.

Status: Common, and found in most areas of the Mediterranean.

Biology: Encountered in depths of 2 m and deeper, either singly or in small very loose groups, favouring sand and mud bottoms, but also found over rocky areas, especially where there are patches of softer sediment nearby. Often present in lagoons and estuaries. Feeds on small fish and crustaceans, worms and molluscs that live on the seabed.

Snorkel tips: A difficult species to spot because of its excellent camouflage, and will often lie semi-buried in sand with just the eyes protruding. Most often seen when it is moving over the seabed. This species can change colour to match that of the seabed. Normally relies on its camouflage to avoid detection and will lie still even when approached close. Although most usually associated with sand and mud, it is quite common to encounter this species resting on flat rocks in mixed seabed environments.

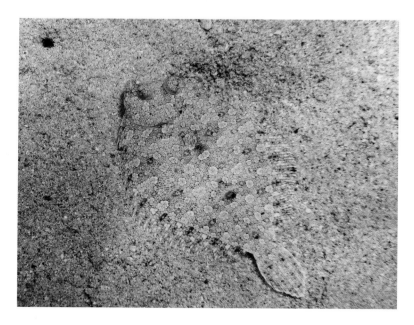

European Flounder

Platichthys flesus

Description: Brown upper-body colour, which can change in shade depending on the background environment. There is usually a scattering of subtle orange spots on the top of the body. Body is oval shaped, and the end of the tail is straight. The eyes are usually on the right-hand side, but left-sided fish are not uncommon. Up to 35 cm in length.

Status: Patchily distributed in the Mediterranean and probably most frequent in northern areas. Also occurs in the Black Sea.

Biology: The European Flounder favours estuaries and brackish-water habitats, and will even travel up rivers into completely fresh water. Most frequently occurs over mud and sand. Frequently buries itself into the sediment so that just the eyes are showing. In winter, will often migrate to deeper water.

Snorkel tips: A familiar fish of the north-east Atlantic, which is locally distributed in the Mediterranean, where it is mostly an estuarine fish, and so is seldom encountered by snorkellers. The cryptic camouflage makes it difficult to spot, but these fish are easy to approach extremely close. If scared, the European Flounder will swim away in an explosive turn of speed.

Surface- and midwater-swimming fish

Blotched Picarel

Spicara maena

Description: Bluish-grey back and silvery-grey sides, with a rectangular dark blotch on the middle of each flank. Large males can have a humpback appearance. There is a single dorsal fin. Can grow up to 20 cm, although those encountered by snorkellers are usually around 12 cm or less in length.

Status: Occurs throughout the Mediterranean, although rarely encountered when snorkelling.

Biology: The Blotched Picarel is an open-water swimming fish that is occasionally encountered in small shoals from about 1 m below the surface over deeper-water areas, such as by rocky headlands or pier heads. Sometimes occurs in the company of other open-water swimmers, such as Pompano (*Trachinotus ovatus*).

Snorkel tips: This is an approachable fish, but its small size and preference for open-water habitats make photography difficult.

Damselfish

Chromis chromis

Description: One of the most easily recognisable fish in the Mediterranean, with its small oval, laterally flattened body, with a deeply forked tail. Dark brown head and body, with obvious, dark-edged scales. Very small young fish are a brilliant iridescent blue. This blue gradually diminishes and totally disappears as the fish mature. Up to 15 cm in length, more typically 8–10 cm.

Status: A common fish that is widely distributed throughout the Mediterranean.

Biology: A shoaling species, often encountered hanging almost stationary in midwater by rockfaces or mixed-bottom seabeds over depths from 2 m to 10 m. Also shoals over seagrass (*Posidonia*) beds. Occasionally, individuals or much smaller groups do occur. The male displays to attract the female, and after mating will guard the eggs, which are laid on rocks. The brilliant-blue young are a familiar sight as they gather in shallow, sheltered areas by rocks. Feeds on zooplankton.

Snorkel tips: Damselfish are ubiquitous and very likely to be encountered when snorkelling. They are confiding fish and it is usually possible to swim through a shoal and cause minimal disturbance. They often like to shoal by steep rockfaces. In shallow water, look out for the electric-blue juveniles, which look so different from the adults that it is hard to believe that they are the same species.

Garfish

(Garpike, Sea Needle)

Belone belone

Description: A fast, silvery, surface swimmer, with an elongated, thin body and long beak-like jaws. Up to 80 cm in length, more usually around 30–40 cm.

Status: Occurs throughout the Mediterranean.

Biology: Usually seen swimming just below the surface, singly, in small groups or in larger shoals. Prefers to swim over deeper water, but will often approach close to steep rockfaces, or by harbour breakwaters. Feeds on smaller fish and will sometimes jump out of the water when feeding or escaping predators. Garfish tend to spend the winter offshore and only approach the coast during summer.

Snorkel tips: An unmistakable fish, given its peculiar shape and long 'beak'. The less frequent Atlantic Saury or Skipper (*Scomberesox saurus*) is very similar, the main difference being the presence of a series of finlets behind the dorsal and anal fins, although these are very difficult to see. Garfish are wary and tend to keep a reasonable distance from snorkellers. They can be easily missed because they swim so close to the surface and a snorkeller needs to be looking ahead, rather than downwards, in order to spot one.

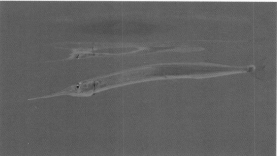

Greater Amberjack

(Greater Yellowtail)

Seriola dumerili

Description: A powerful surface and midwater swimming fish. Body is slender, silvery-grey and tinted with yellow. A darkish band runs from the shoulder and through the eye to the upper jaw. Young fish show much yellow and the flanks are barred. Length can reach over 100 cm, although those seen by snorkellers are usually much smaller.

Status: Found throughout the Mediterranean, although often localised. Probably more frequent in eastern parts.

Biology: A fast-swimming surface to midwater fish that occurs in small groups of usually between two and ten individuals, sometimes more. Hunts as a pack, feeding upon smaller pelagic fish such as sardines. Young fish are often found under floating logs, seaweed masses or other such objects.

Snorkel tips: The best chance for seeing these rapid swimmers is to snorkel over deeper areas or steeply shelving banks and by rocky headlands. When feeding, they go into a fast-moving frenzy as they target a shoal of small fish. Greater Amberjack are often curious and will circle a snorkeller briefly on first encounter, although will maintain a reasonable distance. It is always worth checking any floating object, as the young like to congregate beneath. The author has even found young fish sheltering under a floating black plastic bag!

Pompano

Trachinotus ovatus

Description: A silvery, oval-shaped fish with distinctive, deeply forked tail. Black tips to the tail fins and second dorsal and anal fins. There are dark spots on the flanks of larger fish. Can reach lengths of 30 cm or more, but those seen by snorkellers are usually 10–14 cm.

Status: Occurs widely throughout the region.

Biology: A fast-swimming open-water fish that prefer areas with disturbed water such as around rocky headlands and the surge zone in shallow water by pebble beaches. Most often seen in small groups.

Snorkel tips: Pompano are quick-moving fish that are not often seen, but they are loyal to certain sites such as close to the surf zone of beaches and can be observed repeatedly in such areas. They are easy to approach close, but their speed of swimming and small size makes them tricky to photograph.

Yellowmouth Barracuda

Sphyraena viridensis

Description: Distinctive long, slender body shape, with pointed snout. Numerous dark vertical bars down the silvery-grey flanks. The notably forked tail is often yellowish and edged black. Up to 100 cm in length, but more typically in the region of 30–40 cm. This species can easily be confused with the similar looking European Barracuda (*Sphyraena sphyraena*).

Status: Found throughout the Mediterranean and often encountered by snorkellers.

Biology: Occurs most frequently in shoals of 10 or more individuals, swimming in midwater or just below the surface. This species is a fast and powerful swimmer, feeding on pelagic fish such as sardines and horse mackerel.

Snorkel tips: Barracudas often have a bad reputation, but snorkellers need not fear this species, for those in the Mediterranean are harmless to humans and feed only on fish. However, the Yellowmouth Barracuda is a curious species and a shoal will often repeatedly swim around a snorkeller, providing a good opportunity to take photographs. To encounter this fish, the best technique is to snorkel on the surface over water deeper than 5 m. Rocky headlands and close to harbour walls are often good places to seek them out. Yellowmouth Barracudas often have favoured spots and it is possible to observe the same shoal in the same area day after day.

Lessepsian migrants

A significant number of fish species native to the Red Sea, and which are part of the Indo-Pacific fauna, have colonised the eastern and parts of the central Mediterranean through the artificial conduit of the Suez Canal. Known as Lessepsian migrants, many have a detrimental impact on the environment here, upsetting its delicate ecological balance. The Rivulated and Squaretail Rabbitfish (*Siganus rivulatus* and *S. luridus*) are the two species most likely to be encountered by snorkellers. Cyprus, Turkey, Rhodes and other parts of the eastern Aegean are where you are most likely to see such Lessepsian migrants, although many are spreading their range westwards all the time.

Rivulated Rabbitfish

(Marbled Spinefoot)

Siganus rivulatus

Description: Silvery-grey flanks with a subtle, yellowish tinge on the upper body. Superficially similar to the Salema Bream (*Sarpa salpa*) in shape and colouration. Faint, thin yellow lines run along the flanks. Tail is slightly forked (unlike in Squaretail Rabbitfish) and there are robust spines on the dorsal and anal fins. Length 18–24 cm.

Status: A coloniser from the Red Sea, first recorded in the Mediterranean in 1927 and now found in many eastern parts (common in Cretan waters) and at least as far west as the Ionian Islands.

Biology: Found in small to medium-sized shoals over rocky bottoms from 2 m depth. Occasionally will group into larger shoals. Feeds mainly on algae and is likely to be serious competitor with the native Salema Bream.

Snorkel tips: Rivulated Rabbitfish can be quite common by rocky shores in the Eastern Mediterranean. They are relatively fast swimming, but are tolerant of snorkellers and will allow a reasonably close approach.

Squaretail Rabbitfish

(Dusky Spinefoot)

Siganus luridus

Description: Heavier in build than the Rivulated Rabbitfish (*Siganus rivulatus*). Body is greenish, although colour and patterning can vary considerably, and often the fish has a two-toned appearance of darker above and paler below. The tail edge is straight. Typical length 14–24 cm.

Status: An immigrant from the Red Sea, the Squaretail Rabbitfish is established in many parts of the Eastern Mediterranean and is locally common.

Biology: Most frequently occurs over shallow rocky and mixed-habitat seabeds, usually in small groups but sometimes singly. Feeds on algae. Often erects its venomous spiny fins, which act as a means of protection against predators.

Snorkel tips: Squaretail Rabbitfish are confiding and easy to approach close. They often like shallow water by rocky headlands where there is a strong sea surge.

Other Lessepsian migrants

Common Lionfish (*Pterois miles*) is a spectacular species, with long, splayed pectoral fins, and a prominently spined dorsal fin. The body is heavily vertically striped with shades of beige and dull pink. It often hovers just above the seabed. The lionfish has a painful, dangerous sting and this is a fish that should not be approached too close or ever handled.

The **Red Squirrelfish** (*Sargocentron rubrum*), also known as the **Red Sea Striped Squirrelfish**, is a nocturnal species that can be found in caves and dark recesses during the day.

The **Red Sea Goatfish** (*Parupeneus forsskali*) is similar in shape to the Red Mullet, but is pale coloured with a prominent dark horizontal stripe along the flank and a yellow tail. Found over soft bottoms. The **Goldband Goatfish** (*Upeneus moluccensis*) has a longitudinal golden stripe along the side.

Red Sea Goatfish

Sharks and rays

Sharks and rays are characterised by having cartilaginous skeletons, and a series of gill slits. The fins are fleshy and stiff, and not supported by fine bony spines as in other fish.

It is unusual to encounter a shark or ray when snorkelling due to their scarcity and preference for deeper-water areas. For safety reasons, it would be wise not to approach too close to a shark, and caution should also be applied to rays, some of which have the capability to sting or give an electric shock. There follows a summary of some species that may be encountered in shallow waters – however, it should be remembered that there are a good number of other different shark and ray species in the Mediterranean, including the **Great White Shark** (*Carcharodon carcharias*) – although this species is very rare and tends to live offshore.

Blue Shark (*Prionace glauca*) is occasionally recorded inshore, in shallow-water areas. It is an elegant, slim shark with a blueish body and paler underparts, growing up to 3 m in length.

Blue Shark

The **Sand Tiger Shark** (*Carcharias taurus*) is more robust than the Blue Shark and sometimes ventures into inshore areas, swimming close to the seabed. The **Sandbar Shark** (*Carcharhinus plumbeus*) also explores shallow seabeds on rare occasions. They have been recorded near fish farms off the Turkish coast.

Sand Tiger Shark

Sandbar Shark

The **Small-spotted Catshark** (*Scyliorhinus canicula*) is a small shark, growing up to 70 cm, with a multitude of small dark spots scattered across its sandy-coloured body. Also known as the Lesser Spotted Dogfish, this is a bottom-living shark that feeds on crustaceans, molluscs and small fish.

For rays, the **Common Stingray** (*Dasyatis pastinaca*) is an occasional visitor to shallow-water areas – beware of its dangerous sting. The **Common Torpedo Ray** (*Torpedo torpedo*) is another infrequent venturer close to shore, preferring soft-sediment seabeds. This species can give an electric shock.

Common Stingray

Torpedo Ray

OTHER VERTEBRATES

Loggerhead Turtle

Caretta caretta

Description: A large turtle, usually 100–130 cm in length, with a red-brown to greenish shell, light-coloured limbs and a large, mottled head.

Status: Loggerheads can be found throughout the Mediterranean, although they are uncommon, and mainly breed in eastern half of the basin, with the principal egg-laying sites located in Greece, Turkey, the south of Italy, Cyprus and parts of the North African coast. Turtles from the Atlantic also enter the Mediterranean, although most are visitors and do not breed.

Biology: Loggerheads approach close to shore from May until early September, where they will lay their eggs in holes dug in sandy beaches at night. The holes are then covered in sand. Once hatched, the young quickly make their way to the sea. Loggerheads feed on molluscs, crustaceans and jellyfish.

Snorkel tips: Encountering a Loggerhead Turtle when snorkelling is an unusual but memorable experience. They are vulnerable to tourist disturbance near their nesting beaches and the best ploy is to give one a wide berth if encountered and leave it in peace. The much scarcer Green Turtle (*Chelonia mydas*) occurs in the Eastern Mediterranean, especially around Turkey and Cyprus.

A marked Loggerhead Turtle nesting site with buried eggs in Greece.

A Loggerhead nesting beach in Kefalonia, Greece.

Bottlenose Dolphin

Tursiops truncatus

Description: A robust, large dolphin, grey on the upperside and paler beneath. The 'beak' is short and stubby, the forehead rounded. Up to 3.5 m in length.

Status: Found throughout the Mediterranean in low numbers and may be declining. The bottlenoses in the Mediterranean show genetic differences from those in other areas of the world.

Biology: Feeds on fish, squid and crustaceans and occurs in both coastal and offshore areas. In the Mediterranean, occurs in groups of usually fewer than 10 individuals.

Snorkel tips: More likely to be found near to the coast than other dolphin species, and occasionally approaches close to swimmers – although one would need to be very lucky indeed to encounter these animals when snorkelling. The **Short-beaked Common Dolphin** (*Delphinus delphis*) also occurs in the Mediterranean but is more offshore in its habits. However, if you are visiting the Mediterranean region, both species are worth looking out for when on boat trips.

Bottlenose Dolphin

Short-beaked Common Dolphins

European Shag

(Mediterranean Shag)

Gulosus aristotelis desmarestii

Description: Similar in appearance to a Cormorant (*Phalocrocorax carbo*), but smaller and with a more slender bill. Colour varies depending on age and the breeding cycle, but ranges from black (with a green, iridescent sheen) to a dull grey-green. When breeding there is a small crest on top of the head.

Status: Common throughout Mediterranean – Shags found in the Mediterranean are a subspecies of the European Shag.

Biology: Found along rocky coastlines and often seen perched on rocks, holding the wings open. Dives from the surface to feed on a wide variety of small, bottom-dwelling fish.

Snorkel tips: This is the most commonly encountered seabird when snorkelling, and occasionally one sees Shags swimming under the water, using their wings to propel themselves forward at great speed. When perched on rocks, it is usually possible for a snorkeller to approach close without unduly alarming the bird.

Little Egret
Egretta garzetta

Description: A distinctive, white-plumaged heron-like bird, with dark beak and legs, and yellow feet. Long, feathery streamers on head and breast.

Status: Found throughout the Mediterranean.

Biology: While most often associated with lagoons, rivers, lakes and wetlands, Little Egrets also haunt a variety of sheltered shores in the Mediterranean, especially in winter and early spring. Feeds on a variety of small fish and crustaceans.

Snorkel tips: Little Egrets are occasionally seen by snorkellers, stalking the shore edge for fish and other prey. By taking a careful, slow approach when in the water, it is sometimes possible to get close to one without it taking flight.

Grey Heron

Ardea cinerea

Description: A large heron, with grey wings, a long white neck that is dark-streaked on the front, and a white head topped with black. There are long, feathery streamers on the breast. Length: 90–98 cm.

Status: Found locally on Mediterranean coasts.

Biology: This species breeds near some coastal areas of the Mediterranean, but is probably most likely to be encountered by snorkellers during the migration season in spring and autumn, as well as in winter. On the coast, the Grey Heron feeds mainly on fish and crustaceans.

Snorkel tips: An unmistakable large bird that may occasionally be encountered by snorkellers when it fishes along rocky shores or by coastal wetlands. There is a distinct migratory movement of Grey Heron through Malta in spring and autumn. Generally a shy bird and hard to approach close, but it is a great species to observe when hunting. The flight is slow and lumbering.

Yellow-legged Gull

Larus michahellis

Description: Very similar to the Herring Gull of Northern Europe, the Yellow-legged Gull has only recently been recognised as distinct species. The legs are yellow, and the grey on the wings and the back of the adult is slightly darker than in the Herring Gull.

Status: Widespread and common throughout the Mediterranean.

Biology: Found along all types of coastlines and often occurs in harbours, and coastal towns and cities. An opportunist, it eats a wide range of food items, including fish, insects, birds, worms and human garbage.

Snorkel tips: Often seen by snorkellers in the sea or flying overhead. This gull is usually quite approachable.

Lesser Black-backed Gull

Larus fuscus

Description: Similar in shape and appearance to Yellow-legged Gull (*Larus michahellis*), but with darker, charcoal-grey wings and back. Length: 52–67 cm.

Status: A passage migrant (autumn and spring) and winter visitor to the Mediterranean, especially western parts.

Biology: An opportunist, feeding on a wide range of items, including fish, carrion and invertebrates. Breeds in Northern Europe, and moves south to spend the winter, favouring the West African coast.

Snorkel tips: Occasionally seen while snorkelling during winter and when on migration in autumn and early spring. Most likely to be encountered on the Mediterranean southwestern coast of Spain near Gibraltar.

Audouin's Gull

Ichthyaetus audouinii

Description: A beautiful, elegant gull with a distinctive, dark red beak with a black and yellow tip. Paler and narrower winged than Yellow-legged Gull.

Status: Localised distribution, breeding mainly in Spain (including Balearic Islands), Italy and Greece. Population is thought to be decreasing.

Biology: Breeds on small islands, either in colonies or alone. Tends to be more of an open-sea feeder than Yellow-legged Gull, preying upon small surfacing-swimming fish such as sardines and anchovies. However, the author has also seen this species scavenging on food dropped by tourists on beaches.

Snorkel tips: Not often seen because of their relative scarcity – but well worth looking out for when snorkelling in places such as mainland Spain, Ibiza, Mallorca, Menorca, Corsica and Sardinia, as well as the Aegean Sea.

Black-headed Gull

Larus ridibundus

Description: An elegant gull, with slender wings, pale grey upperparts and white below. In summer, the head is chocolate brown (appears black from a distance), and in winter the head is white with a black spot behind the eye. The bill is slender. Length: 35–38 cm.

Status: Found locally on the Mediterranean coast during winter, early spring and autumn.

Biology: A colonial breeder, nesting on freshwater marshes and other wetlands, with some birds from northern parts of Europe migrating south to spend winter in the Mediterranean. It can often be seen hawking for flying insects.

Snorkel tips: Not usually encountered during the summer, but sometimes seen when snorkelling during the autumn or winter, especially December to February. Often likes to congregate on harbour breakwaters. Can be confused with the slightly larger Mediterranean Gull (*Larus melanocephalus*), which has a shorter, heavier beak and a jet-black head in spring and summer.

Common Sandpiper

Actitis hypoleucos

Description: An energetic little wading bird with short legs, grey upper plumage and breast, and white belly. Length: 19–21 cm.

Status: A passage migrant (appearing in autumn and spring) in coastal areas of the Mediterranean.

Biology: Feeds on a range of aquatic invertebrates and nests on the ground by fresh water. Breeds in Northern Europe and winters in Africa, passing across the Mediterranean on migration.

Snorkel tips: Common Sandpipers are frequently encountered by snorkellers in rocky areas of the Mediterranean when on migration in late summer and early autumn. These are restless birds, forever on the move, and the body often bobs up and down as they pause.

Kingfisher

Alcedo atthis

Description: An unmistakable small bird, with azure-blue upperparts and orange breast. Dumpy in the appearance, the Kingfisher has a long, straight bill. Length: 16–17 cm.

Status: Breeds on rivers and lakes, but in autumn and winter often moves to the Mediterranean coast.

Biology: Specialises in hunting small fish caught from a perch from which the bird dives into the water. When on the coast, prefers sheltered rocky areas where there are calm areas of sea, enabling favourable conditions for catching smelts and other small fish, as well as prawns.

Snorkel tips: Kingfishers are occasionally seen when snorkelling later in the year, towards autumn and into winter, most usually as a flash of electric-blue as the bird flies low over the water on fast-whirring wings. Often utters a high-pitched cry as it speeds by.

INVERTEBRATES

The diversity of marine invertebrates is one of the highlights when snorkelling in the Mediterranean. Identification down to species level can, however, often be difficult under snorkel conditions.

Molluscs

Molluscs are a large group of animals that includes gastropods (for example sea hares), bivalves (such as mussels and clams) and cephalopods (for instance octopus and cuttlefish). Many possess a hard outer shell to protect their bodies, although the shell has become lost or internalised in other species.

Common Cuttlefish

Sepia officinalis

Description: Colour varies depending on background, but usually brownish with paler stripes along the back, giving a zebra-like appearance. Has a flattened 'skirt' around the bottom fringe of the body. The two eyes are prominent, and the tentacles point forward from the large head. Typical length: 20–25 cm.

Status: Occurs throughout the Mediterranean, although it is not often encountered when snorkelling.

Biology: Common Cuttlefish are probably most likely to be seen in spring and early summer when they come inshore to breed – although the author has encountered them in shallow water in Spain in autumn. They can change colour quickly to match their background. Cuttlefish are predators, feeding on a range of prey including crustaceans and fish.

Snorkel tips: Cuttlefish are fascinating creatures that can be encountered singly, in pairs or even in small groups. Their behaviour can be unpredictable: sometimes they are shy of an approaching snorkeller, whereas at other times they might be confiding and allow a close approach. On occasion it is possible to witness a cuttlefish catching prey, involving a careful stalk before quickly extending its tentacles to seize the victim. Look out too for another species, the Elegant Cuttlefish (*Sepia elegans*), which is smaller and slenderer than the Common Cuttlefish and lacks stripes.

Common Octopus

Octopus vulgaris

Description: Body length about 1 m, including arms. Main part to the body oval shaped, with two large distinctive and well-developed eyes, and siphon often visible. The eight arms have suckers on the undersides. Grey is the most common background colour, with darker mottling, but can change shape and colouration to match the background environment.

Status: Common throughout Mediterranean, especially over rocky ground.

Biology: Occurs in water as shallow as 1 metre, but usually deeper. Often seen squeezed tight into a rock crevice or a gap under a boulder. Although largely nocturnal, octopus can also be encountered hunting over the seabed during the day for prey such crustaceans and other shellfish. Most usually found over rocky bottoms but also sometimes over sand and mud.

Snorkel tips: The Common Octopus is fickle and can be abundant in some areas and scarce in others. Their nocturnal habits mean that the most productive time to find them is just after daybreak. The best way to seek them out is to snorkel in water less than 5 m deep and regularly check under gaps in rocks and boulders on the seabed. Octopus are faithful to their dens, and will occupy the same site for prolonged periods, making individuals easy to find on subsequent snorkels, once initially located. Shellfish fragments are frequently encountered at the mouth of a den. When caught out in open water and panicked, an octopus will squirt a dark cloud of ink as a defensive ploy. Octopus are intelligent creatures and are sometimes curious of snorkellers, especially if the approach is slow and there are no sharp movements.

Spotted Sea Hare
Aplysia dactylomela

Description: A colourful sea slug, ranging from yellow to brown or green and with dark-rimmed rings covering the fleshy body. There are two pairs of tentacles (rhinophores) on the head, the front pair resembling hare's ears. Length 7–25 cm, although the body can shrink itself to squeeze into crevices or when danger threatens. There are wing-like lobes along the side of the body.

Status: Most frequent in eastern and central parts of the Mediterranean, and in the Adriatic.

Biology: A recent coloniser to the Mediterranean from the Atlantic, it has spread rapidly since 2002, possibly due to warming sea temperatures. Can occur in large numbers in spring and early summer when these animals gather to breed, often in sheltered creeks and inlets. Found in shallow water from 1 m downwards, usually where there are rocks. Also occurs in seagrass (*Posidonia*) meadows. The Spotted Sea Hare is herbivorous, grazing on algae as it crawls across rocks or the seabed. The rhinophores on the head are used to detect chemicals present in the water, including pheromones from other sea hares.

Aplysia dactylomela

Snorkel tips: The Spotted Sea Hare is a most striking gastropod mollusc that is easy to approach close. Keep an eye out for this species clinging to rock shelves in shallow water. Another much smaller species of Sea Hare worth looking for in shallow, sheltered areas is **Elysia timida**, which is cream/white coloured, speckled with dark spots, and only about 3 cm in length.

Elysia timida

Sea Hare

(Spotted Sea Hare)

Aplysia punctata

Description: Colour varies from olive-green to reddish-brown to black, often with small light spots on the body. There are four tentacles (rhinophores) on the head. Usually 10–20 cm in length. Lobed wings on the side of the body.

Status: Occurs in many parts of the Mediterranean, especially Spain and the northern Adriatic.

Biology: Found over a variety of bottoms, the Sea Hare is a gastropod mollusc, but has no external shell. When threatened by a predator the animal releases a purple cloud of ink combined with a milky mix of chemicals that confuses the attacker. The fleshy 'wings' are used to aid propulsion over the seabed. Sea hares may change colour to match the surrounding environment and their diet may also influence colour. Sea Hares are hermaphrodite (each animal is both male and female) and often form mating chains, with some animals acting as females, others as males, and some as both. The Sea Hare will often move into shallower water in spring and summer to breed, when large numbers may congregate.

Snorkel tips: Smaller than the Spotted Sea Hare (*Aplysia dactylomela*), this species is also more unobtrusive. It is much commoner in northern Atlantic waters than it is in the Mediterranean. It is easy to approach close.

Giant Doris
Felimare picta

Description: An attractive sea slug, with a grey-blue body and yellow markings. Grows up to 20 cm in length.

Status: Found throughout the Mediterranean, but infrequently encountered when snorkelling.

Biology: Found from about 3 m and deeper over rocky seabeds. They are hermaphrodites and feed on sponges.

Snorkel tips: This is a stunning nudibranch that is easy to approach close, although in most instances one needs to dive under because they prefer to avoid very shallow water.

Other molluscs

The **Atlantic Triton Shell** (*Charonia variegata*) is an attractively patterned, large predatory gastropod mollusc. The author has once encountered this species in Crete.

With a distinctive long, narrow protuberance at the front end, the empty shells of the **Spiny Dye Murex** (*Bolinus brandaris*) are often found over shallow rocky and mixed bottoms when snorkelling. The species is sometimes also known as the **Purple Dye Murex** because in ancient times this mollusc was a source of dye. They are also commercially harvested and eaten as food.

Ceriths (Cerithidae) are distinctive, narrow conical gastropod molluscs, the empty shells of which are often seen when snorkelling. A good majority of such shells are frequently occupied by hermit crabs. The most commonly encountered type is the **Common Cerith** or **Horn Shell** (*Cerithium vulgatum*), which grows up to 7 cm long. Also occurring in the Mediterranean are the smaller and very similar looking *C. lividulum* and *C. renovatum*.

Limpets are frequent on rocks by the splash line. Several species occur in the Mediterranean, including Mediterranean Limpet (*Patella caerulea*), Lusitanian Limpet (*Patella rustica*) and Rough Limpet (*Patella ulyssiponensis*). On soft rock, limpets can leave little circular depressions on the surface caused by their incredibly strong rasping teeth. Limpet teeth are constructed from one the strongest materials yet discovered in the natural world, rivalling even the toughest human-made equivalents.

The attractive, spiral-shaped empty shells of **Spindle Euthria** (*Euthria cornea*) are sometimes encountered when snorkelling.

The **Sicilian** or **Syracusan Spindle Shell** (*Aptyxis syracusanus*) is an elongated gastropod with attractive shell colouring. Grows to about 5 cm in length.

The shells of the **Banded Dye Murex** (*Hexaplex trunculus*) are often spotted when snorkelling, and are frequently occupied by hermit crabs.

The **Turbinate Monodont** or **Turban Shell** (*Phorcus turbinatus*) is commonly encountered on rocks in shallow water down to a few metres depth. The spiralled shell is attractively patterned with spots.

The **Mediterranean Mussel** (*Mytilus galloprovincialis*) is very similar and closely related to the familiar Common Mussel (*Mytilus edulis*) which is widely found in the North Atlantic. Smooth dark shell, length up to 10 cm. There are often other organisms growing on the shell, including barnacles and the white calcareous tubes of keel worms. Found throughout the Mediterranean, although the distribution is patchy, given this species' dependence upon specific habitat requirements. Lives in shallow, intertidal waters, most especially areas protected from the open sea such as lagoons, harbours, inlets and estuaries. Attaches to rock and other hard substrates by byssus threads and may occur on soft bottoms. Often occurs in large clusters, for example on wooden or metal pier supports. This species filter feeds on plankton and spawns from September to May. The larvae settle after only a few weeks, attaching themselves to a suitable substrate.

The **Date Mussel** (*Lithophaga lithophaga*) is a commonly occurring elongated mussel, which bores into carbonate rocks in shallow water. They are very slow growing and can live for a long time.

The **Noble Pen Shell** or **Fan Mussel** (*Pinna nobilis*) is the Mediterranean's largest mollusc, growing up to 1 m in length! Lives upright with the narrow, pointed end secured to the seabed by byssus threads. The outside of the shell has a rough appearance due to the growth of algae, tubeworms, sponges and other organisms. Scarce and patchily distributed throughout the Mediterranean and declining due to overfishing for food and the decorative properties of the shell, as well as from disturbance by boat anchors. It is also highly susceptible to a parasite that has become prevalent in recent years, causing widespread mortality. Classed as 'Critically Endangered' on the IUCN Red List of Threatened Species. This is a sedentary mollusc that mostly occurs at depths below 5 m. Most usually found in seagrass (*Posidonia*) beds, but also sometimes on mix rocky and sandy bottoms. When filter feeding, the two valves (or shells) of the mussel open. Seldom seen by snorkellers due to its scarcity and tendency to live in deeper water. The thick growth of organisms on the outer shell makes the Noble Pen Shell difficult to spot from the surface. This is a species that requires diving down to see close-up, but do not touch because of its threatened status and vulnerability.

Worm Snails or **Worm Shells** (family Vermetidae) occur in cemented clusters. They have a curled, segmented appearance with an obvious body-opening like a snail. Light grey in appearance. They are found growing on rocks in the splash zone and shallow water. Apart from the free-floating larval stage, they are sessile creatures that resemble hardened coiled worms when seen close, or barnacles from a distance. They feed on suspended particulates in seawater. Worm Shells or Worm Snails are easy to overlook as their thick growths often resemble grey rock. It is always worth taking the time to examine rocks closely for such creatures. When visibility is poor, for example when the water has recently been turbulent, closely inspecting rock surfaces is a great way of seeing life when it is harder to spot more obvious creatures such as fish.

Echinoderms

Echinoderms include such well-known creatures as starfish, sea urchins and sea cucumbers. They are characterised by their radial symmetry, with their body parts extending outwards from the mouth (although this is not apparent in sea cucumbers). The name echinoderm means 'spiny skin', and while this is usually the case, not all echinoderms have this feature. They move by using their tube feet, which are operated by a water vascular system.

Red Starfish

(Red Sea Star)

Echinaster sepositus

Description: A medium-sized bright orange to red starfish. The five (occasionally up to seven) arms taper gently to blunt tips. The upper surface is rough textured and the span across the body and arms is typically 10–20 cm.

Status: Common throughout the region.

Biology: Lives in shallow water, usually on rocky bottoms, but also mud, sand and seagrass (*Posidonia*) beds where it lies with arms outstretched. Feeds on detritus, algae, sponges and small invertebrates.

Snorkel tips: This is the commonest starfish in the Mediterranean and the one most likely to be encountered by snorkellers. Body has a slightly slimy, soapy feel. This species could easily be confused with the Purple Starfish (*Ophidiaster ophidianus*), but it is brighter and more orange in colour, with a rougher skin texture.

Purple Starfish
(Purple Sea Star)

Ophidiaster ophidianus

Description: A large, attractive starfish, with relatively smooth skin. Usually a deep cherry to almost purple in colour, but also exhibits other shades of red. May have pale mottled spots. Up to 25 cm in diameter, occasionally larger.

Status: Frequent throughout the Mediterranean, especially in central parts.

Biology: Found in water from about 2 m and deeper, most usually on rocky bottoms. Feeds mainly on coralline algae, but also small sponges and other invertebrates.

Snorkel tips: A striking species, and deeper red in colouration than the superficially similar Red Starfish (*Echinaster sepositus*). As well as on the seabed, look out for Purple Starfish clinging to underwater rockfaces.

Spiny Starfish
Marthasterias glacialis

Description: A distinctive and often large starfish, with five tapering arms. The upper surface is covered in short, thick spines, set in three longitudinal rows. The central disc of the body appears small in relation to the length of the arms. Colour varies, most typically is blue-grey or greenish but also sometimes brown and even purple. Usually 20–30 cm across but can grow much larger.

Status: Most frequently encountered in the Eastern Mediterranean and Adriatic. Also found around the Balearic Islands and parts of mainland Spain.

Biology: Lives on a variety of bottoms, including rock and mud, in shallow water from about 2 m downwards. Feeds on other starfish, crabs, molluscs and fish carrion. Spawning aggregations may gather in summer in shallow water.

Snorkel tips: The Spiny Starfish is a striking species and unlikely to be confused with other starfish. It is easy to observe close. The best places to search are sheltered, less exposed areas of the seabed.

Blue Spiny Starfish

(White Starfish)

Coscinasterias tenuispina

Description: A small starfish with a variable number of short-spined arms of varying length, typically from six to eight, although sometimes ten arms. There are rows of small, pale spines running down each arm. Size: 10–15 cm across from one arm tip to the other opposite. Tan background colour with spots and a subtle bluish tinge.

Status: Occurs patchily in the Mediterranean, possibly more frequently in central and eastern parts. Can be common in areas where it is found.

Biology: Found in shallow water, frequently resting on rock shelves just below the splash zone. Often shows more movement than other types of starfish. Feeds on a range of invertebrates.

Snorkel tips: As with other starfish, the Blue Spiny Starfish is easy to approach. Look out for them over any areas with rock.

Rock Sea Urchin and Black Sea Urchin

Paracentrotus lividus and *Arbacia lixula*

Description: Distinctive, dark spine-covered body (known as a test) that grows to about 7 cm in diameter. The body of the Rock Sea Urchin (*P. lividus*), also known as Atlantic Purple Sea Urchin, is marginally flatter in shape than the similar Black Sea Urchin (*A. lixula*). Spines up to 3 cm long. The Rock Sea Urchin frequently has a purple or light brown hue, whereas the Black Sea Urchin is black. However, such colour variation is not always obvious, making it challenging to differentiate between these two species under snorkel conditions. The Rock Sea Urchin frequently covers its spines with seaweed and other marine debris.

Status: Both species are common throughout the Mediterranean.

Biology: Found on rocks from very shallow water downwards. Most frequently on sloping and near vertical rocks, but also common on rocky bottoms. The Black Sea Urchin seems to prefer vertical rockfaces, while the Rock Sea Urchin is more prevalent on horizontal or sloping rock shelves. These urchins fix themselves to rocks, sometimes leaving a cavity in the rock wall. They are herbivores, grazing on algae around the favoured resting spot. May be scarce or absent from areas heavily frequented by tourists because the animals have been removed to prevent foot injury.

Snorkel tips: Sea urchins are commonly encountered when snorkelling and you need to take great care not to inadvertently touch one when using hands to propel yourself along rocks, or to step on one when entering or leaving the water. The brittle spines can break off once inside the skin and easily become infected if not treated. Look out too for the less frequently encountered **Purple Sea Urchin** (*Sphaerechinus granularis*), which is larger and has white tips to its violet spines.

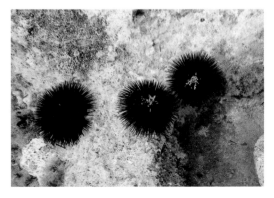

Tubular Sea Cucumber

(Common Sea Cucumber)

Holothuria tubulosa

Description: Long, tubular body Up to 30 cm in length, although more typically 20 cm. The body surface is covered in numerous conical protrusions (papillae). Colour varies greatly from near black to pale brown.

Status: Common in many parts of the Mediterranean, particularly in central and eastern areas.

Biology: One of the commonest sea cucumber species in the Mediterranean Sea. Found over a variety of bottoms including sand, mud, rock and seagrass (*Posidonia*) beds. Moves slowly over the bottom by means of tubular feet. Feeds on organic particles. Often covers its body with detritus, such as dead seaweed.

Snorkel tips: This species is similar in appearance to the closely related Cotton-spinner (*Holothuria forskali*), so called because it can eject defensive white threads from its rear end when the body is touched. These will deliver a mild stinging sensation to the hand.

Variable Sea Cucumber

(Patchy Sea Cucumber)

Holothuria sanctori

Description: A strikingly patterned sea cucumber, with cream spots set against a dark brown body. Each pale spot has a dark centre. Body cylindrical and slug-like, and rounded at either end. The skin is soft. Can grow up to 25 cm.

Status: Found throughout the Mediterranean, possibly more frequent in central and eastern parts.

Biology: Encountered on rocky and mixed seabeds, as well as on rock ledges. Moves along the surface substrate very slowly feeding on detritus, plankton and ingesting sand and mud particles, from which it can extract food items; sea cucumbers are important marine recyclers. There is an endoskeleton just beneath the skin.

Snorkel tips: Sea cucumbers are often encountered by snorkellers and are curious creatures to observe. They are usually most abundant in sheltered locations where there is a lot of detritus. In some parts of the world they are much sought after as food. Look out too for the **White Spot Sea Cucumber** or **Sandy Cucumber** (*Holothuria poli*). It is not nearly as frequently seen as the Variable Sea Cucumber or the Tubular Sea Cucumber (*Holothuria tubulosa*) and is most often found on sandy bottoms and among seagrass (*Posidonia*) beds.

Cnidarians

Cnidarians include sea anemones, jellyfish and corals, and are radially symmetrical. They are typified by their stinging cells used to capture prey. Ctenophores – such as comb jellies – look similar to cnidarians, but lack stinging cells, among other differences.

Mauve Stinger Jellyfish

Pelagia noctiluca

Description: A striking jellyfish with pinkish-purple bell (up to 10 cm circumference) and long trailing stinging tentacles (up to 1 m).

Status: Found throughout the Mediterranean and can occur in huge swarms that are possibly associated with planktonic blooms.

Biology: A near-surface-living jellyfish, which is often found singly but can occur in huge numbers close to the shore, and often associated with other species, such as Compass Jellyfish. May move up and down the water column regularly in an undulating movement.

Snorkel tips: This is a jellyfish to approach with extreme caution as the sting is painful, and it is easy to drift inadvertently in among their long trailing tentacles. The author has been stung on several occasions and the irritation can take up to a month to subside. When snorkelling in areas where Mauve Stingers are prolific, frequently look all around you to prevent coming too close and be aware that the thin trailing tentacles can be difficult to see.

Compass Jellyfish

Chrysaora hysoscella

Description: A large jellyfish with a brownish-yellow bell that reaches up to 25 cm in diameter. Prominent dark lines radiate down from the apex at the top of the bell. The tentacles are long and hang from the rim.

Status: Occurs throughout the region.

Biology: Most frequent in spring, when large aggregations can occur. This species changes sex, starting off as male and later turning into a female.

Snorkel tips: Always give Compass Jellyfish a wide berth as they have a nasty, painful sting. The tentacles can trail quite far back, so always be aware of this.

Crystal Jellyfish

(Many-ribbed Jellyfish)

Aequorea forskalea

Description: A beautiful, bioluminescent creature, with a relatively flat bell fringed with short tentacles, some longer than others. The transparent bell has many radial canals. Diameter of bell typically 15–17 cm.

Status: Found throughout the Mediterranean.

Biology: Sometimes great blooms of this species occur, when many animals can be encountered. Feeds on zooplankton and other tiny organisms.

Snorkel tips: Crystal jellies are easy to approach close and are mesmerising to watch as they gently pulsate, often moving up and down in the water column. Storm surges can cause large numbers to accumulate in rocky creeks, gullies and bays. Although the tentacles have stinging cells, they are not powerful enough to damage human skin and so do not pose any significant threat.

Other Jellyfish and jellyfish-like animals

The **Barrel Jellyfish** (*Rhizostoma pulmo*) is an unmistakable species. This massive cream-coloured jellyfish really can grow to the size of a barrel, but they are gentle giants, feeding on plankton and their sting is mild and not generally harmful to humans. The biggest specimens can exceed 1.5 m in length and 35 kg in weight.

The **Moon Jellyfish** (*Aurelia aurita*) is a disc-shaped translucent jellyfish through which are visible four distinctive, purplish rings which make up the reproductive organs. The bell is surrounded by small tentacles, which are not harmful. This species often occurs in large swarms.

Keep an eye out too for **Comb Jellies**. Although they look like jellyfish, they actually belong to their own phylum, the Ctenophora. They are translucent, typically oval in shape, and have eight rows of 'combs' which run down the body. On the combs are tiny hairs or 'cilia' that help provide movement, and which through the resultant action of light scattering deliver a rainbow effect, making these extremely attractive creatures. They do not sting.

Sea Gooseberry
Pleurobrachia pileus

Description: A species of comb jelly, the Sea Gooseberry is oval-shaped, with two long feathery tentacles, which can be completely retracted but which can be up to 10 cm long. Body (without tentacles) is about 2–3 cm long. Tiny iridescent hairs, or cilia, used for propulsion, run down the body in eight rows.

Status: Found throughout the Mediterranean.

Biology: The two tentacles are used to catch prey. They have 'sticky' cells which small food items adhere to.

Snorkel tips: While typically an offshore, pelagic species, good numbers can sometimes be encountered in inshore waters. The shimmering movement of the cilia creates a wonderful rainbow light effect. This species reaches its highest densities in summer. Very challenging to photograph because of their small size, and semi-transparent body.

Portuguese Man o' War

Physalia physalis

Description: Although similar in appearance to a jellyfish, the Portuguese Man o' War is a siphonophore, comprising a mutually dependent colony of animals that work together as one. It has a distinctive, purple-pink translucent floatation bladder, with trailing tentacles beneath. The length of the bag is typically around 18–20 cm, although often smaller.

Status: More usually found in subtropical parts of the Atlantic, it infrequently occurs in the western Mediterranean where large swarms can occasionally appear.

Biology: The Portuguese Man o' War is a floating colony of different zooids that have specialised functions such as catching prey, forming the float, or for reproduction and digestion. The long and poisonous trailing tentacles stun and kill prey such as fish and surface crustaceans.

Snorkel tips: This is a creature to avoid when snorkelling because its long tentacles have an exceptionally painful sting, which in extreme cases can be fatal to humans. The gas floatation balloon is conspicuous on the surface. It is always good practice when snorkelling to scan the surrounding area above the water every so often for orientation purposes and to check for potential hazards such as this creature or approaching boats.

By-the-wind Sailor

(Purple Sail)

Velella velella

Description: With the circumference of a large coin, this most unusual looking organism floats upon the sea surface. Set upon a bright blue disc rises a semi-transparent convex 'sail'. Beneath the disc are small tentacles. Diameter: 3–5 cm.

Status: Found throughout the Mediterranean, most commonly in western parts, including the Ligurian Sea and around the Balearic Islands where huge aggregations sometimes occur in late spring and summer.

Biology: The By-the-Wind sailor is an ocean drifter carried by the vagaries of wind and current. In zoological parlance it is termed a colonial hydrozoan: each individual is a colony of specialised polyps, similar in fashion to the more familiar Portuguese Man o' War (*Physalia physalis*). The tentacles beneath the disc are used to catch plankton.

Snorkel tips: By-the-Wind Sailors are chance encounters for snorkellers, but when they do occur, it is often in vast numbers. During such occasions, winds and currents push them into rocky creeks and inlets where they collect in large aggregations. This frequently happens in conjunction with massive jellyfish swarms, which occur more frequently in the Mediterranean nowadays, perhaps in response to changes in sea temperature and the occurrence of planktonic blooms.

Beadlet Anemone

Actinia equina

Description: A distinctive bright red anemone, up to 4 cm in diameter and 6 cm high. When seen close, a blueish circumference ring may be discerned at the bottom of the anemone, and blue bead-like spots (acrorhagi) encircle of the top of the body just below the tentacles. When feeding, the short tentacles are displayed. The tentacles are retracted when out of the water, giving the animal the appearance of a cherry-red, jellified blob.

Status: Common throughout the Mediterranean.

Biology: A shallow-living species that occurs on rocks from the narrow Mediterranean intertidal zone down to around 3 m depth. Most often found on rockfaces just under the surface, secreted in fissures and hollows, sometimes singly but often in loose groups. Often encountered below rock underhangs, especially on those that are exposed at low tide. Despite appearing sedentary, these anemones are territorial and will use their tentacles to drive away neighbouring anemones that encroach upon their space.

Snorkel tips: Some careful searching of vertical and sloping rock outcrops will often bear fruit in encountering this species. The tentacles contain stinging cells (nematocysts), which act just like microscopic harpoons that inject small planktonic prey with poison. Don't worry if you accidently brush against one, as the stinging cells of the Beadlet Anemone are indetectable to humans.

Snakelocks Anemone

(Green Anemone)

Anemonia viridis

Description: A pale green to greyish (often beige) anemone, with long tentacles that are sometimes tipped with purple. Including the tentacles, the body span is up to 15 cm across. Due to their density and length, the tentacles are unable to fully retract.

Status: Common throughout the region.

Biology: Occurs in shallow water in rocky areas, frequently on rockfaces and shelves, and on large rocks on seabed. Also attaches itself to seagrass. Feeds on small fish and invertebrates that are trapped by its tentacles. The anemone contains symbiotic algae, which are necessary for it to thrive. The algae require light to photosynthesise, and for this reason the Snakelocks Anemone occurs only in shallow water.

Snorkel tips: Often encountered by snorkellers in rocky habitat in the water less than 5 m deep. Distinctive and easy to identify, although the colour can vary, but usually a pale green, verging on beige or grey. Take care as the tentacles have the potential to sting.

Trumpet Anemone

Aiptasia mutabilis

Description: A green to light brown, short-tentacled anemone, with a slender column that is narrower at the base and wider at the top, although in most instances this is hard to discern. The body is typically small and about 5 cm across. The tentacles taper to a point and are slightly flattened and semi-transparent.

Status: Found in all parts of Mediterranean, but not nearly as frequently encountered as Beadlet or Snakelocks Anemones.

Biology: Lives on rocks in shallow water, often just beneath the surface. Often encountered in clusters growing in gentle depressions on rockfaces and ledges.

Snorkel tips: The Trumpet Anemone is an easy species to overlook because it often clings close to rockfaces with the column of the body hidden in shallow indentations and pock marks, or by calcified seaweeds and other encrustations. Close examination of rockfaces and ledges in shallow water is required. The translucent tentacles are attractive and sometimes have a subtle blueish tint.

Girdle Anemone

Actinia cari

Description: A brown to beige anemone with distinctive dark concentric lines encircling the body column. Short tentacles that can be fully retracted. When the tentacles are retracted, the body has a conical appearance. Up to 4 cm wide and 6 cm high.

Status: Found in all parts of Mediterranean, but uncommon.

Biology: Lives in shallow water, often on rocky outcrops just under the surface, frequently sheltering beneath rock ledges. Mainly feeds at night with tentacles usually retracted during the day.

Snorkel tips: This species is seldom encountered by snorkellers, but it is probably overlooked because of its relatively inconspicuous nature, frequently living beneath rock underhangs just below the surface. The fact that it is mainly nocturnal, and the tentacles are retracted during the day, further adds to the low profile of the animal.

Star Coral

Astroides calycularis

Description: A striking, small orange or yellow cup coral supported by a calcified cup, which is often hard to discern. Anemone-like, short tentacles brim out over the top of the stumpy body column.

Status: Some reference guides cite it as being most frequent in western parts of the basin, although the author has found this species in central and eastern parts as well.

Biology: Cup corals are fascinating animals, and Star Coral is a colonial species consisting of individual polyps that cluster together under rock overhangs and cave entrances from depths of about 2 m and greater. They are set within a calcified cup, which they can withdraw into. The tentacles are used to trap planktonic organisms.

Snorkel tips: Star Coral can be found beneath rock overhangs, often near the surface, and most especially on rockfaces that rise from deep water. Despite the vibrant orange and yellow colouration, they can be hard to spot, so it is always worthwhile to meticulously examine the places where they are most likely to occur.

Look out too for **Pig-tooth Coral** (*Balanophyllia europaea*) – a solitary hard stony coral that is easily overlooked as it blends in well with the rock it attaches to. Resembling a molar tooth, it is greyish in colour and about 1.5–2 cm long. It holds symbiotic algae, and has small, transparent tentacles.

Crustaceans

Crustaceans are a broad group that includes crabs, lobsters, isopods, prawns, shrimps and – perhaps surprisingly – barnacles. They have a hard exoskeleton, which is formed of chitin and provides armour-like protection. The body is composed of three segments: head, thorax, abdomen.

Common Prawn

Palaemon serratus

Description: A small, translucent prawn, the body streaked with thin, dark brown vertical stripes. There are long thin antennae on the head. The first two pairs of legs have tiny claws. When seen up close, the legs may exhibit purple, maroon and yellow markings. Body length varies between 3 cm and 8 cm, occasionally up to 11 cm.

Status: Common throughout the Mediterranean.

Biology: Frequent on rocky and pebbly bottoms, often in very shallow water less than 1 m deep. Also in seagrass (*Posidonia*) beds and areas of soft sediment, can tolerate brackish water. Usually found singly, although sometimes in small groups. Feeds on both algae and tiny invertebrates. The Common Prawn is an important source of food for fish and other marine predators.

Snorkel tips: It is always worth snorkelling in very shallow water close to the shore edge, and this is one of the species most likely to be encountered. They are usually easy to approach close if no sudden movements are made and will often face-up to a snorkeller. If panicked, a prawn will flee with a quick jerk of the body, but normally settles back down again close-by. When seen very close, they exhibit attractive markings.

Saint Piran's Hermit Crab

Clibanarius erythropus

Description: A small hermit crab (approximately 1 cm long) that lives in requisitioned empty gastropod mollusc shells.

Status: Common throughout coastal areas of Mediterranean.

Biology: Found in very shallow water (down to 3 m) on algae-covered rocks, or areas where detritus such as dead seaweed has accumulated. Hermit crabs differ

from other crabs in that they lack a hard external shell. The body is soft and has a sideways spiral that enables it to slip into the empty coiled shell of a gastropod mollusc. A variety of shells are used, but one of the most frequent is the Common Cerith (*Cerithium vulgatum*). There are several different species of hermit crab in the Mediterranean. They feed on algae and other organic debris, as well as small invertebrates.

Snorkel tips: Hermit crabs are frustrating to observe because they remain so well hidden within their requisitioned shells, with only the eyes, tiny pincers and legs revealed. They are wary creatures and will fully withdraw into their shell if a threat is perceived.

Marbled Crab

(Marbled Rock Crab)

Pachygrapsus marmoratus

Description: A small crab, with carapace width up to 5 cm. When seen out of water the body appears dark green, whilst under the water it has a more grey, marbled appearance. The back is noticeably square shaped.

Status: Common throughout the region.

Biology: Frequently seen running about on rocks very close to the shore edge and also just under the water by the splash zone, typically on rockfaces or boulders. Feeds on algae and small invertebrates.

Snorkel tips: This is a comical species to watch, often seen running on flat rock surfaces right by the water's edge and scuttling into the sea if disturbed. It is easier to approach when under the water, but because it most usually occurs just beneath the surface in a few centimetres of water, this species is often missed by snorkellers.

Sally Lightfoot Crab

(Nimble Spray Crab)

Percnon gibbesi

Description: A long-legged crab with yellow bands on the leg joints, and a flattened body. Overall colour a mix of red and blue. The carapace (body) is 2–6 cm wide. The pincers are small.

Status: Found in most parts of the Mediterranean.

Biology: A new arrival to the Mediterranean, first recorded near Sicily in 1999, but now widespread. It may have naturally colonised the Mediterranean from the Atlantic, but it is probable that this process has been aided by ship-borne means, in ballast tanks. Occurs on large boulders in shallow water, usually near some type of shelter or crevice into which it can retreat. Often found by harbour walls and sea protection boulder barriers. A herbivorous crab that grazes mainly on algae. Its spread through the Mediterranean has been remarkable and may be due to the absence of other herbivorous crabs in the region, enabling it to find a productive niche where there are no competitors.

Snorkel tips: This is a commonly encountered crab but is hard to approach close, as it is shy and nimble and quickly dashes under cover if danger threatens. However, often it will only retreat halfway into a crevice or hole. This species definitely requires the slow and cautious approach. If one is sighted, drift up to it gently and it may be possible to get reasonably close.

Warty Crab
Eriphia verrucosa

Description: A stocky, heavily built crab with large black pincers. The legs and claws are often covered in a fine down-like material. Greenish in colour and carapace is up to 8 cm wide, occasionally larger.

Status: Found throughout the Mediterranean.

Biology: A robust, green-coloured crab that is found in shallow water in rocky and mixed seabed habitats, especially areas with plenty of seaweed. Mainly active at night and at dawn and dusk. Feeds on molluscs, worms and other invertebrates. Migrates to shallow water in late spring to breed.

Snorkel tips: The Warty Crab is a nimble and shy species that is hard to approach close, usually quickly fleeing into a crevice to hide if it perceives danger.

Lesser Spider Crab

Maja crispata

Description: A small spider crab with long legs and claws. The pincers on the claws are small. The carapace is narrower at front and broader at rear, giving an overall triangular-shaped back. Up to 6.5 cm long. The body and legs are almost always covered in a thick growth of algae. Can be easily confused with the larger Spiny Spider Crab (*Maja squinado*).

Status: Common throughout the region.

Biology: Found over rocks and mixed bottoms from very shallow water and deeper, typically places where there is good algal cover and plenty of crevices to hide in.

Snorkel tips: The Lesser Spider Crab is really a master of concealment due to its habit of covering itself with a thick, 'furry' coat of algae and other detritus. This species is often encountered when snorkelling and will quickly scuttle to shelter if disturbed.

Barnacles

Several species of barnacle are found in the Mediterranean, including *Chthamalus montagui*, *C. stellatus*, *Amphibalanus amphitrite*, *Balanus trigonus* and *Perforatus perforatus*. Although they look like molluscs, these are in fact complex crustaceans.

The free-swimming larvae moult several times before sinking to a rock or other structure where they are attracted by the presence of other barnacles. Once settled, they cement themselves to the substrate using their antennae. The newly rested barnacles moult once more, creating a final body structure of chalky fused plates, with a hatch-door opening at the top.

Millions of barnacles can occur along a kilometre of typical rocky shore and they are one of the most ubiquitous animals found on our coasts. When feeding, a barnacle opens the central plates on the top of its volcano-shaped shell and uses its modified legs or 'cirri' to repeatedly claw at the water like a grasping hand reaching out for tiny floating organisms.

Isopods

Anilocra species

Anilocra physodes and Anilocra frontalis are small parasitic isopods, about 3 cm in length, with a flattened, segmented, greyish body, and looking similar to a wood-louse. Found throughout the Mediterranean, they are parasites on fish such as wrasse, Salema (*Sarpa salpa*) and other sea breams. They can injure or kill their hosts and are often seen by snorkellers attached to the sides of fish.

Parasitic isopods attach themselves to fish such as breams.

Another isopod worth looking out for is the **Italian Sea Slater** (*Ligia italica*), which resembles a tiny woodlouse, and is commonly found throughout the Mediterranean on rocks just above sea level.

Ascidians

Ascidians (Tunicates) – or **Sea Squirts** – are infrequently found when snorkelling because they are often inconspicuous and generally prefer deeper-water areas. They are so called because they can squirt water if removed from the sea. They most frequently have ovular, sac-like bodies with two siphon openings. Water is drawn into one siphon, and food particles are trapped as the flow passes through the body before exiting via the other siphon. Oxygen is also drawn from the water during this process. Pictured here is the **Pleated Sea Squirt** (*Styela plicata*) growing on the mooring rope of a harbour pontoon in Greece.

An interesting Ascidian is the colonial type from the family **Botrylloides**, which are very small and easy to miss when snorkelling. They can be found growing on a variety of surfaces, including under mooring pontoons in harbours. The species here is probably *Botrylloides leachii*. Such is the small size of these organisms that the author has often only noticed their presence when examining photographs he had taken of other, larger sessile invertebrates found on rocks or other surfaces!

Worms

The most typical group are the segmented worms (annelids) and include lugworms, tubeworms and fireworms. Flatworms belong to a different group – known as the platyhelminths – which are simple, soft-bodied and usually flattened invertebrates.

Bearded Fireworm

Hermodice carunculata

Description: A spectacular bristleworm similar in appearance to a centipede, with a long, laterally compressed and segmented body with numerous white bristle-like appendages, which superficially look like legs. Body is blue-grey, sometimes greenish. The white bristles are often tinged with red. Can grow up to 30 cm, but usually 10–18 cm.

Status: Occurs throughout the Mediterranean, most frequently in central and eastern parts.

Biology: Found in shallow water over a variety of seabed habitats, either individually or in large aggregations. A predatory animal, feeding on a variety of invertebrates, including small crustaceans. Will also scavenge upon dead creatures.

Snorkel tips: Bearded Fireworms are oblivious to snorkellers and easy to approach close. Take special care not to touch this species – the silky white bristles are sharp and can cause painful irritation to the skin.

Tubeworms

Protula tubularia and *Protula intestinum*

Description: It is challenging to tell these two species apart, especially since the colour on their feathery crowns (branchiae) is variable, ranging from red to white. Tubeworms dwell within calcareous tubes, which can reach up to 20 cm in length, although only the small segment poking above the substrate is usually visible.

Status: Widespread throughout the Mediterranean, but never abundant.

Biology: Tubeworms are very colourful animals, with their feather-like branchiae often seen protruding from crevices in rocks. These are sedentary, preferring shady areas of the seabed, including by crevices, cave entrances and overhangs. Occur in depths of 2 m and greater.

Snorkel tips: These amazing animals are always exciting to find, but a very slow approach is needed, as they will quickly withdraw into their calcareous tubes if danger threatens. Other tubeworms likely to be encountered include *Sabella pavonina* and *S. spallazanii* (**Mediterranean Fanworm**), which are bigger than the above, and feature large crowns of feathery branchiae of varying shades of brown and other colours, including cream. Their exact species identification is difficult, and they occur among rocks or in sand and stones that are often covered with small algae.

It is similarly challenging to identify precisely *Branchiomma* and *Acromegalomma* species, even from photographs. The specimen photographed here was found by the author growing in abundance on mooring ropes in a harbour in Kefalonia, Greece. It is housed in a short tube, and the feathery branchiae (tentacles) alternate in coloured horizontal lines of beige brown and dirty white. It may possibly be the **Rock Peacock Worm** (*Branchiomma bombyx*), but it is impossible to determine with certainty.

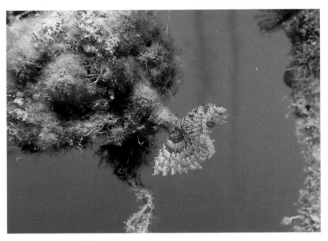

The Terebellidae family of polychaete worms – often known as **Spaghetti Worms** – live in in small burrows or crevices, and have extremely long tentacles used for collecting food particles. The tentacles are fine and hard to see, and this is a worm that needs careful searching along the surface of rocky outcrops.

Other worm-like species

Pseudobiceros splendidus is an attractive flatworm that is occasionally encountered clinging to rocks in shallow water. Oval-shaped and coloured black with golden-yellow on its crinkled margins, it is about 4 cm long.

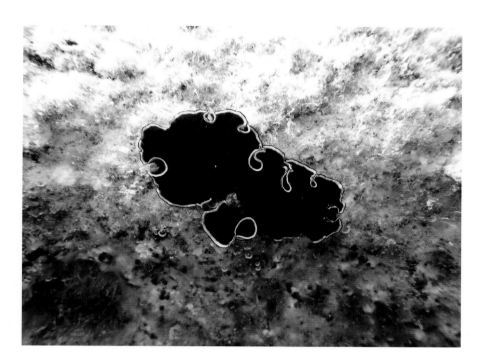

Sponges

Sponges are primitive organisms that are fixed to rocks and seaweeds, and cannot move. They have no sensory organs. Sponges feed and gain oxygen by passing water through their bodies via pores on the surface.

The **Black Leather Sponge** (*Sarcotragus spinosulus*) is a black sponge patterned with pores, 5–18 cm across and encountered throughout the region. A common sponge, often found in shallow water on rocky seabeds, or attached to rock cliffs and shelves. The free-swimming larva settles down onto a substrate where it grows into a young sponge. This species is commonly seen when snorkelling. Most usually found on shallow depressions in rockfaces, in water less than 5 m deep.

The **Yellow Tube Sponge** (*Aplysina aerophoba*) is a distinctive, sulphur-yellow, tubular sponge with a large pore at the end of each projection. Can form colonies up to 1 m across, and is most typically found on rocky (although also sometimes soft sediment) or mixed seabeds in depths of 3 m and greater.

Chondrilla nucula – sometimes known as the **Chicken-liver Sponge** or **Potato Sponge** – is one of the commonest sponges encountered by snorkellers, often growing on rock shelves and underhangs in relatively shallow water. This species prefers open-water areas rather than sheltered places.

It is always worthwhile examining rock surfaces closely, and sponge species that may be found are the **Yellow** or **Sulphur Boring Sponge** (*Cliona celata*) and the **Green Boring Sponge** (*Cliona viridis*). They make holes in soft rocks and only the tiny orange-yellow papillae show above the surface, as the rest of the body is hidden beneath.

Yellow Boring Sponge

Green Boring Sponge

The **Oyster Sponge** (*Crambe crambe*) is a striking species with its orange-red colouration. Also known as the Orange-Red Encrusting Sponge, it occurs on rock-faces from about 2 m and deeper.

The **Golf Ball Sponge** (*Tethya citrina*) is a small, distinctive rotund sponge, orange-red in colour and with a spiky surface that often has silt and other debris attached. Frequently found in rock crevices.

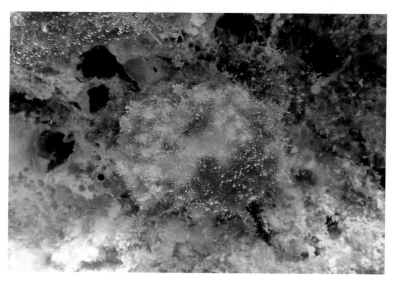

The **Kidney Sponge** (*Chondrosia reniformis*) is frequently encountered by snorkellers on rockfaces, especially beneath underhangs. It has a rubbery appearance and is mottled grey. As the name indicates, this sponge is kidney-shaped, and often occurs in conglomerations.

Bryozoans

Animals belonging to the phylum Bryozoa are fascinating tiny colonial lifeforms called zooids, each one living in a box-shaped cell or compartment, which aggregate to work together as one. They can take on an incredible variety of shapes and forms, including calcified-like encrusting sheets and fans.

The *Schizomavella* and *Schizoporella* species are encrusting bryozoans of different forms that are difficult to identify with certainty. They have hard external surfaces and some – such as *Schizomavella mamillata* – exhibit striking red-orange colouration.

Orange Sea Mat (*Schizomavella mamillata*)

Branching Bryozoan (*Schizoporella errata*), here gaining tenure on a mooring rope.

SEAGRASSES

Although it appears like seaweed, seagrass is an underwater flowering plant rooted to soft sediment on the seabed. Seagrass meadows form one of the most important habitats in the region, providing food and shelter for a wide variety of organisms, including seahorses, pipefish, sea breams, molluscs and bryozoa. These meadows act as crucial nursery areas for fish. Seagrass often grows in dense masses, although sometimes isolated plants are found in sandy areas around rocks. Seagrass meadows act as significant carbon sinks and can in fact absorb carbon dioxide much faster than tropical forests. The meadows also protect coasts from erosion, acting as buffer zones.

Seagrass

(Neptune Grass, Mediterranean Tapeweed)

Posidonia oceanica

Description: The plant arises from creeping stems called rhizomes, which taper into groups of flattened green leaves that can be up to up to 1 m in length but are typically shorter.

Status: Common throughout the Mediterranean, but declining and vulnerable to high seawater temperatures, pollution and disturbance caused by the anchors of boats. The shallow strait between the islands of Ibiza and Formentera in the Balearics is renowned for its seagrass meadows.

Biology: Occurs in shallow water from about 1 m, growing on sand and other soft-sediment bottoms. Usually propagates by vegetative reproduction through the rhizomes, although also sometimes from green flowers that emerge in autumn, with the floating fruits (sea olives) released from May to July. Flowering does not occur every year. Most usually found in depths of 3–15 m. This species is endemic to the Mediterranean.

Snorkel tips: The beauty of seagrass meadows lies in the abundant life held within, so it is always worth diving down to investigate. Look closely for a variety of fish hiding there, including Annular Sea Bream (*Diplodus annularis*) and Deep-snouted Pipefish (*Syngnathus typhle rondeleti*). Epiphytic communities of bacteria, algae and bryozoa colonise the surface of the leaves.

Decaying seagrass washed up on a beach.

Broadleaf Seagrass (*Halophila stipulacea*) is an invasive species of seagrass from the Red Sea that is encountered in shallow, sheltered areas of the Eastern Mediterranean.

SEAWEEDS

Seaweed is the general term for a large group marine algae that come in a variety of different forms, shapes and colours. They are classified into three broad groups based on their colour: brown, red and green, which are referred to as Phaeophyceae, Rhodophyceae and Chlorophyceae, respectively. Seaweeds are often ignored by snorkellers because of their ubiquity – but look closely and their colour and diversity will never fail to enthral.

Examine rocks faces closely to see many with intricate forms, and overhangs and the entrances to sea caves where pink encrusting coralline species may be found. The scope and variety of seaweed is immense, which often makes species identification challenging.

The red **Asparagopsis** algae – *Asparagopsis taxiformis* and *A. armata* – are attractive red or red-pink seaweeds, which are non-native colonisers that have spread widely throughout the Mediterranean. They originate from tropical to warm temperate areas of the Atlantic and Indo-Pacific (in the case of *A. taxiformis*), or from Southern Ocean temperate areas around Australia and New Zealand (*A. armata*). Both species are considered to cause ecological harm to the Mediterranean basin by eroding native species biodiversity. *A. armata* is commonly known as Red Harpoon Weed. Found growing in shallow water.

Common Coral Weed (*Corallina officinalis*) is a pinkish seaweed that exhibits a rough skeletal texture. This species is common in rocky areas of the Mediterranean and is similar in appearance to Elongate Coral Weed (*C. elongata*), which has longer, feather-like fronds.

Jania rubens is a pink-hued seaweed that can cover large expanses of rock in shallow areas, often having the appearance of fluffy pink balls.

The ochre-coloured **Laurencia obtusa** – or Rounded Brittle Fern-weed – is an attractive red alga found growing on rocks in shallow water.

Dictyota – sometimes known as fan weed – are attractive brown algae with delicate fronds. Species in the Mediterranean include *D. dichotoma*, *D. spiralis* and *D. fasciola*.

Flabellia petiolata is a common green seaweed with small, flattened leaves that usually grows in shady positions, such as under rocks.

Halimeda tuna is a distinctive small seaweed with flattened, disc-shaped fronds that are held vertically aloft. It is often found in shady places among rocks.

Sea Lettuce (*Ulva* species) are vibrant, limey-green seaweeds that occur in sheets on the seabed in shallow, sheltered and sunny locations. Often prefer nutrient-rich areas.

Colpomenia are unusually shaped, spherical seaweeds sometimes encountered on rocks in shallow areas of the Mediterranean. Known as the **Oyster Thief**, two similar species occur: *C. sinuosa* and *C. peregrina*.

Peacock's Tail (*Padina pavonica*) *is a* distinctive dirty-white seaweed, with a calcareous parchment-like texture and fronds shaped into a shallow funnel. Around 3–5 cm across at the widest point, occasionally larger. Peacock's Tail is ubiquitous in many shallow areas of the Mediterranean close to the shore, abounding on rocks, especially in sheltered, sunny areas. Rarely found in water deeper than 5 m. It is worth closely investigating the fronds closely, as small gobies, blennies and crustaceans frequently use this seaweed for shelter.

With the delightful name of **Mermaid's Wine Glass**, the *Acetabularia* group of seaweeds are most attractive, with their pale, shallow, cup-shaped discs held aloft on thin spindly stems. They are commonly found growing in very shallow water, often on mooring ropes. These are among some of the largest known single-celled organisms.

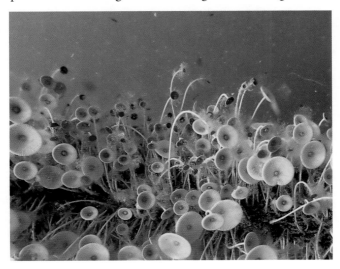

Amphiroa rigida is a blue-grey, calcified, branched seaweed often found attached to rock shelves and rockfaces in shallow water, especially in sheltered areas.

The calcified pink branches of ***Tricleocarpa fragilis*** are a real eyecatcher, growing on rockfaces and ledges from about 2 m depth. Most likely to be found in the Eastern Mediterranean.

Titanoderma trochanter is an attractive, grey to beige coralline, spherical alga found in eastern parts of the Mediterranean, often on rock shelves from 1 m depth.

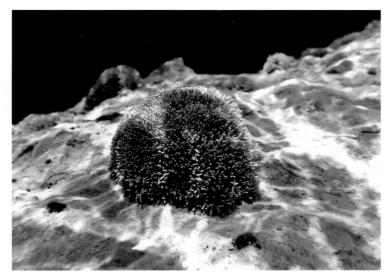

Sea Rose (*Peyssonnelia* species) is frequently encountered in low-light environments such as sea caves and rock underhangs from 0.5 m in depth. Brownish-red in colour, these disc-shaped algae form into tiers. *Peyssonnelia squamaria* and *P. rosa-marina* both occur in the region.

The **Pink Paint Weeds** (Corallinaceae) are striking coralline red seaweeds (including *Lithothamnion* species) that grow as calcified encrustations over rocks, bringing real colour and vibrancy. Some species are smooth, others rough and knobbly. Despite being common, individual species are extremely hard to identify.

PHOTOGRAPHIC CREDITS

INDEX